Library Handbook for
Organic Chemists

Library Handbook for Organic Chemists

Andrew J. Poss
Honeywell International, Inc.
Buffalo, New York

Chemical Publishing Co., Inc.
New York

"Two weeks in the laboratory will save you three hours in the library."

From Professor Samuel Danishefsky's lecture on Quadrone at the University of Rochester.

CONTENTS

Preface   xi

Acknowledgments   xiii

A. Search Orientation Table   xv

B. Primary Sources

    B.1. Aldrich Catalog/Handbook of Fine Chemicals   1

    B.2. Annual Reports in Organic Synthesis   5

    B.3. Beilstein's Handbook of Organic Chemistry   11
       B.3.a. Overview   11
       B.3.b. Distribution of Compound Types   16
       B.3.c. Abbreviations   20
       B.3.d. Rules of the Beilstein System of Indexing   25
       B.3.e. List of Prefixes   28

    B.4. Chemical Abstracts   51
       B.4.a. Overview   51
       B.4.b. Roles   57
       B.4.c. Online File Names   59
       B.4.d. Patent Country Codes   72
       B.4.e. Chemical Abstracts Sections   74

    B.5. The Chemist's Companion   83
       B.5.a. Overview   83
       B.5.b. Index to the Table of Contents   84

    B.6. Compendium of Organic Synthetic Methods   91

    B.7. Comprehensive Organic Chemistry   95
       B.7.a. Overview   95
       B.7.b. Index to the Table of Contents   96

B.8.   Comprehensive Organometallic Chemistry   99
       B.8.a. Overview   99
       B.8.b. Index to the Table of Contents   100

B.9.   Comprehensive Organic Functional Group
       Transformations   107

B.10. Comprehensive Organic Transformations   109

B.11. CRC Handbook of Chemistry and Physics   111
       B.11.a. Overview   111
       B.11.b. Index to the Tables   113

B.12. CRC Handbook of Data on Organic Compounds   137

B.13. Dictionary of Organometallic Compounds   143

B.14. Dictionary of Organic Compounds (Heilbron)   147

B.15. Dissertations   151

B.16. Encyclopedia of Reagents for Organic Synthesis   153
       B.16.a. Overview   153
       B.16.b. Journal Abbreviations   154

B.17. Formation of C–C Bonds (Mathieu)   159
       B.17.a. Overview   159
       B.17.b. Index to the Table of Contents   160

B.18. Houben-Weyl Methoden der Organischen Chemie   167
       B.18.a. Overview   167
       B.18.b. Index to the Table of Contents   169

B.19. Internet Resources   179

B.20. Journals   181

B.21. Kirk-Othmer Encyclopedia of Chemical Technology   187

B.22. Lange's Handbook of Chemistry   189
       B.22.a. Overview   189
       B.22.b. Index to the Sections   190

B.23. Literature Sources   215

B.24. The Merck Index   217
       B.24.a. Overview   217
       B.24.b. Alphabetical List of Tables   221

B.25. Organic Reactions   223

B.26. Organic Synthesis   225

B.27. Patai's Chemistry of Functional Groups Series   229
    B.27.a. Overview   229
    B.27.b. Index of Titles   231

B.28. Patents   243

B.29. Purchasing Chemicals   249

B.30. Reagents for Organic Synthesis (Fieser's)   251

B.31. Rodd's Chemistry of Carbon Compounds   253
    B.31.a. Overview   253
    B.31.b. Index to the Table of Contents   256

B.32. Science Citation Index   283

B.33. Scientific Dictionaries   287

B.34. Theilheimer's Synthetic Methods of Organic   289
    Chemistry
    B.34.a. Overview   289
    B.34.b. Index to the Systematic Classification   293
    B.34.c. Functional Group Transformations   303

B.35. Translations   305

C. Organic Chemistry Reviews   309

Index to Reviews in *Tetrahedron* and *Synthesis*   311

Index   355

PREFACE

Library Handbook for Organic Chemists is an effort to bridge the gap between the needs of the research scientist and the predicaments of today's libraries. Science libraries vary in collection size, completeness, and budget. Many libraries have incomplete sets of the large expensive series such as Beilstein's, Houben-Weyl, or Rodd's. These collections are often missing important indexes and are thus rendered useless to the chemical searcher. Furthermore, it is either too late or still too expensive to complete these series. An incomplete reference collection may hold the answers to a scientist's questions, but they have no means to access the information. This book introduces the Index to the Table of Contents and provides it to as many sources as possible. The Index to the Table of Contents is expected to make incomplete collections useful to researchers. For example, if your library has a partial collection of Houben-Weyl, say Volumes 1–12, you would have to search the index in each volume to determine if the information you seek is available in this collection. The cumulative index was not published until Volume 16. By checking the Index to the Table of Contents in the Houben-Weyl section of this book, you can determine which volume to look up your desired information or borrow that volume by interlibrary loan if it is not readily available.

Throughout the book, Library of Congress call numbers are provided to aid the researcher in finding references. The journal section lists the call numbers for journals because some libraries continue to file scientific journals by call number, not alphabetically by title. (*An excellent way to turn three hours into six hours.*) Information searching is becoming only a mouse click away from the laboratory. Researchers can start their library searches without ever leaving their desk by making use of the Internet web sites and searchable CD-ROMs mentioned throughout the text.

Section A is a link between commonly asked questions or needed information and the chemical library sources available. Section B is an expansion of each secondary library source with suggestions for easy access, indexes available, organization, and indexes to their table of contents to make the user more familiar with the resource. Section C is an index of all review articles published in *Tetrahedron* and *Synthesis* journals. Although these reviews can be accessed through other reference materials, they constitute many of the concepts of organic chemistry and quick access to them could lead the searcher to his quest faster than other reference sources.

This book is expected to streamline and organize the process of searching the chemical literature. It represents an effort to ensure that the correct reference works and all possible sources are consulted during a search. For those using computer searching, it is expected to help in selecting databases and to aid in searching. This book should be the first reference source a researcher uses when beginning a literature search and the last book he or she consults near its completion. It is written and organized so that a searcher does not have to read the entire book or even an entire section to understand and search the chemical literature. Each section of the book is designed for quick assimilation of the information necessary to find a reference.

Andrew J. Poss
Buffalo, N.Y., 2000

ACKNOWLEDGMENTS

I thank Mrs. Janice Hood, librarian at Honeywell International, Inc., Buffalo, New York; Mrs. Pia Moller-Reibsch, Bibliothek at Honeywell International, Inc., Seelze, Germany; and Mrs. Pam Santorelli, librarian at OxyChem, Grand Island, New York.

I express my deepest appreciation to "Beany Mae" for her support and encouragement during the preparation of this text.

I also acknowledge the publishers and database companies for their permission to reproduce information found in this book.

Company	Section
Academic Press, Inc.	Annual Reports in Organic Synthesis
Aldrich Chemical Co., Inc.	Aldrich Catalog/Handbook of Fine Chemicals
American Chemical Society	Chemical Abstracts; CRC Handbook of Data on Organic Compounds; Dissertation Abstracts; Merck Index; Patents; Science Citation Index; Theilheimer's Synthetic Methods of Organic Chemistry
Beilstein-Institut	Beilstein's Handbuch der Organischen Chemie
CRC Press	CRC Handbook of Chemistry and Physics; CRC Handbook of Data on Organic Compounds
The Dialog Corporation	Dictionary of Organometallic Compounds; Dictionary of Organic Compounds

Elsevier Science	Comprehensive Organometallic Chemistry; Comprehensive Organic Chemistry; Rodd's Chemistry of Carbon Compounds
Institute for Scientific Information	Science Citation Index
John Wiley & Sons, Inc.	Chemist's Companion; Compendium of Organic Synthetic Methods;
John Wiley & Sons Limited	Encyclopedia of Reagents for Organic Synthesis
Karger, Basel	Theilheimer's Synthetic Methods of Organic Chemistry
Kluwer Academic Publishers	Dictionary of Organometallic Compounds; Dictionary of Organic Compounds
The McGraw-Hill Companies	Lange's Handbook of Chemistry
Merck & Co., Inc.	Merck Index
Thieme, New York	Formation of C–C Bonds; Houben-Weyl

Search Orientation Table

To Find ▼	See ▼
Analysis–classes of compounds	Houben-Weyl, p. 167 Patai's, p. 229 Rodd's, p. 253
Analytical methods	Beilstein, p. 11 Chemical Abstracts, p. 51 Houben-Weyl, p. 167 Patai's, p. 229
Article, citations of	Science Citation Index, p. 283
Articles by author	Science Citation Index, p. 283 Chemical Abstracts, p. 51
Author's papers	Science Citation Index, p. 283 Chemical Abstracts, p. 51
Catalysts	Comp Organometallic Chem, p. 99 Chemical Abstracts, p. 51
Chemical derivatives	Beilstein's, p. 11 Patai's, p. 229 Chemical Abstracts, p. 51 CRC Handbook, p. 111 Dictionary Org Compds, p. 147 Merck Index, p. 217
Chemical names	Aldrich, p. 1 Merck Index, p. 217 Chemical Abstracts, p. 51 CRC Org Compds, p. 137 Dictionary Org Compds, p. 147 Lange's Handbook, p. 189
Chemical procurement	*Purchasing Chemicals, p. 249*
Chemical properties	CRC Org Compds, p. 137 Chemical Abstracts, p. 51 Kirk-Othmer, p. 187 Rodd's, p. 253 Houben-Weyl, p. 167 Dictionary of Org Compds, p. 147 Beilstein's, p. 11
Chemical subjects, general	Kirk-Othmer, p. 187

To Find ▼	See ▼
Chemicals, purchase	*Purchasing Chemicals, p. 249*
Citations of papers	Science Citation Index, p. 283
Classes of compounds—analysis	Houben-Weyl, p. 167 Rodd's, p. 253 Comprehensive Org Chem, p. 95 Chemical Abstracts, p. 51
Classes of compounds—preparation	Comprehensive Org Chem, p. 95 Houben-Weyl, p. 167 Kirk-Othmer, p. 187 Rodd's, p. 253 Chemical Abstracts, p. 51
Classes of compounds—reactions	Comprehensive Org Chem, p. 95 Houben-Weyl, p. 167 Kirk-Othmer, p. 187 Rodd's, p. 253 Chemical Abstracts, p. 51
Compound, classes of—analysis	Houben-Weyl, p. 167 Rodd's, p. 253 Chemical Abstracts, p. 51
Compound, classes of—preparation	Comprehensive Org Chem, p. 95 Houben-Weyl, p. 167 Kirk-Othmer, p. 187 Rodd's, p. 253 Chemical Abstracts, p. 51
Compound, classes of—reactions	Comprehensive Org Chem, p. 95 Houben-Weyl, p. 167 Kirk-Othmer, p. 187 Rodd's, p. 253
Compound preparation	Beilstein, p. 11 Dictionary Org Compds, p. 147 Chemical Abstracts, p. 51 En. Reag. Org. Syn, p. 153. Kirk-Othmer, p. 187 Merck Index, p. 217 Organic Synthesis, p. 225 Fieser, p. 251 Rodd's, p. 253

To Find ▼	See ▼
Compound reference	Aldrich, p. 1 Beilstein's, p. 11 Chemical Abstracts, p. 51 Merck Index, p. 217 CRC Org Compds, p. 137 Dictionary Org Compds, p. 147 Lange's Handbook, p. 189
Compound structure	Aldrich, p. 1 Beilstein's, p. 11 Chemical Abstracts, p. 51 CRC Org Compds, p. 137 Dictionary Org Compds, p. 147 Lange's Handbook, p. 189 Merck Index, p. 217
Compounds, industrial uses	Kirk-Othmer, p. 187
Compounds, reactions of	En. Reag. Org. Syn, p. 153. Chemical Abstracts, p. 51 Beilstein's, p. 11 Organic Synthesis, p. 225 Form C-C Bonds, p. 159
Conversion tables	Chemist's Companion, p. 83 CRC Handbook, p. 111 Lange's Handbook, p. 189 Merck Index, p. 217
Conversions, synthetic	Ann Reports Org Syn, p. 5 Patai's, p. 229 Chemical Abstracts, p. 51 Theilheimer, p. 289 Compendium, p. 91 Comp Org Fn Gr Trans, p. 107 Larock, p. 109 Comprehensive Org Chem, p. 95
Derivatives, chemical	Beilstein's, p. 11 Patai's, p. 229 Chemical Abstracts, p. 51 CRC Org Compds, p. 137 Dictionary Org Compds, p. 147 Merck Index, p. 217

To Find ▼	See ▼
Dictionaries, language	*Translations, p. 305*
Dictionaries, scientific	*Sci. Dic, p. 287*
Di-functional compds, preparation	Compendium, p. 91 Comp Org Fn Gr Trans, p. 107 Larock, p. 109
Dissertations	*Dissertations, p. 151* Chemical Abstracts, p. 51
Experimentals	Houben-Weyl, p. 167 Organic Synthesis, p. 225 Organic Reactions, p. 223
Experimental techniques	Chemist's Companion, p. 83 Organic Synthesis, p. 225 Houben-Weyl, p. 167
Foreign language dictionaries	*Translations, p. 305*
Functional group transformations	Ann Reports Org Syn, p. 5 Patai's, p. 229 Chemical Abstracts, p. 51 Larock, p. 109 Compendium, p. 91 Form C-C Bonds, p. 159 Comp Org Fn Gr Trans, p. 107 Theilheimer, p. 289
Functional groups, protection	Compendium, p. 91 Comp Org Fn Gr Trans, p. 107 Larock, p. 109 Theilheimer, p. 289
General chemical subjects	Kirk-Othmer, p. 187
Industrial process	Kirk-Othmer, p.187
Industrial uses of compounds	Kirk-Othmer, p. 187
Information sources	*Literature Sources, p. 215* *Internet Resources, p. 179*
Internet language dictionaries	*Translations, p. 305*

To Find ▼	See ▼
Internet resources	*Internet Resources, p. 179*
Journal location	*Journals, p. 181* Chemical Abstracts, p. 51 (CASSI)
Language dictionaries	*Translations, p. 305*
Literature sources	*Literature Sources, p. 215*
Mathematical tables	Chemist's Companion, p. 83 CRC Handbook, p. 111 Lange's Handbook, p. 189 Merck Index, p. 217
Names, chemical	Aldrich, p. 1 Beilstein's, p. 11 Chemical Abstracts, p. 51 CRC Handbook, p. 111 Dictionary Org Compds, p. 147 Lange's Handbook, p. 189 Merck Index, p. 217
Name reactions	Merck Index, p. 217 Organic Synthesis, p. 225 *Reviews, p. 309* Houben-Weyl, p. 167
Organometallic compounds	Comp Organometallic Chem, p. 99
Papers, citations of	Science Citation Index, p. 283
Patents	Chemical Abstracts, p. 51 *Patents, p. 243*
Pharmaceutical properties	Merck Index, p. 217 Chemical Abstracts, p. 51
Physical properties	Aldrich, p. 1 Beilstein's, p. 11 Chemical Abstracts, p. 51 Chemist's Companion, p. 83 CRC Org Compds, p. 137 Dictionary Org Compds, p. 147 En. Reag. Org. Syn, p. 153. Lange's Handbook, p. 189 Merck Index, p. 217

To Find ▼	See ▼
Preparation—classes of compounds	Comprehensive Org Chem, p. 95 Houben-Weyl, p. 167 Kirk-Othmer, p. 187 Rodd's, p. 253
Preparation of compounds	Beilstein's, p. 11 Dictionary Org Compds, p. 147 Chemical Abstracts, p. 51 En. Reag. Org. Syn, p. 153. Kirk-Othmer, p. 187 Merck Index, p. 217 Organic Synthesis, p. 225 Fieser's, p. 251 Rodd's, p. 253
Preparation of di-functional compds	Compendium, p. 91 Comp Org Fn Gr Trans, p. 107 Larock, p. 109
Process, industrial	Kirk-Othmer, p. 187
Properties, chemical	Kirk-Othmer, p. 187
Protection of functional groups	Compendium, p. 91 Comp Org Fn Gr Trans, p. 107 Larock, p. 109 Theilheimer, p. 289
Purification of compounds	Fieser's, p. 251 Rodd's, p. 253 Houben-Weyl, p. 167
Reactions—classes of compounds	Comprehensive Org Chem, p. 95 Houben-Weyl, p. 167 Kirk-Othmer, p. 187 Rodd's, p. 253
Reactions, named	Merck Index, p. 217 Organic Synthesis, p. 225 Houben-Weyl, p. 167 *Reviews, p. 309*

To Find ▼	See ▼
Reactions of organic compounds	En. Reag. Org. Syn, p. 153. Form C-C Bonds, p. 159 Theilheimer, p. 289 Chemical Abstracts, p. 51 Beilstein's, p. 11
Rearrangements	Comprehensive Org Chem, p. 95 Houben-Weyl, p. 167 Chemical Abstracts, p. 51 Patai's, p. 229
References, compound	Aldrich, p. 1 Beilstein's, p. 11 Merck Index, p. 217 Chemical Abstracts, p. 51 CRC Org Compds, p. 137 Dictionary Org Compds, p. 147 Lange's Handbook, p. 189
Registry number	Chemical Abstracts, p. 51 Aldrich, p. 1 CRC Org Compds, p. 137 Merck Index, p. 217 Lange's Handbook, p. 189
Review articles	Ann Reports Org Syn, p. 5 Compendium, p. 91 Chemical Abstracts, p. 51 Larock, p. 109 En. Reag. Org. Syn, p. 153. Science Citation Index, p. 283 Theilheimer's, p. 289 Reviews, p. 309
Scientific dictionaries	*Sci. Dic, p. 287*
Sources, information	*Literature Sources, p. 215* *Internet Resources, p. 179*
Sources, internet	*Internet Resources, p. 179*
Spectroscopic references	Aldrich, p. 1 Beilstein's, p. 11 Chemical Abstracts, p. 51 CRC Org Compds, p. 137

To Find ▼	See ▼
Structure, compound	Aldrich, p. 1 Beilstein's, p. 11 Chemical Abstracts, p. 51 CRC Org Compds, p. 137 Dictionary Org Compds, p. 147 Lange's Handbook, p. 189 Merck Index, p. 217
Synthetic conversions	Ann Reports Org Syn, p. 5 Patai's, p. 229 Chemical Abstracts, p. 51 Theilheimer, p. 289
Tables, conversion	Chemist's Companion, p. 83 CRC Handbook, p. 111 Lange's Handbook, p. 189 Merck Index, p. 217
Tables, mathematical	Chemist's Companion, p. 83 CRC Handbook, p. 111 Lange's Handbook, p. 189 Merck Index, p. 217
Techniques, laboratory	Chemist's Companion, p. 83 Organic Synthesis, p. 225 Houben-Weyl, p. 167
Transformations, functional group	Ann Reports Org Syn, p. 5 Patai's, p. 229 Chemical Abstracts, p. 51 Compendium, p. 91 Form C-C Bonds, p. 159 Comp Org Fn Gr Trans, p. 107 Theilheimer's, p. 289
Translations, dictionaries	*Translations, p. 305*

B.1.

Aldrich Catalog/Handbook of Fine Chemicals:

TP202.A38

- **compound name and structure**
- **physical properties**
- **chemical and spectroscopic references**

The Catalog/Handbook contains information on only those chemicals sold by Aldrich Chemical Company.

EASY ACCESS

— For a chemical, look in the Molecular Formula Index for the systematic name and page number.

INDEXING

In the **Chemicals Section**, all chemicals contained in the catalog are listed by a systematic name in alphabetical order. Prefixes, i.e., *bis*, *tris*, *tetra*, etc., are part of the chemical name. Descriptors, i.e., *R*, *S*, *cis*, *trans*, *tert*, etc., are not used in alphabetizing the chemical name. Commonly used names and acronyms are cross-referenced to the systematic name and catalog page where they can be found.

The **Molecular Formula Index** lists all chemicals contained in the Aldrich Catalog by Hill molecular formula. For each formula, the products are arranged in alphabetical order along with the catalog page number where the entry can be found. Formulas are arranged by citing carbons in ascending order, then hydrogens, and finally the remaining elements in alphabetical order. If carbon is not present, the elements are arranged in alphabetical order. Metal salts of acids, alcohols, and amines, and hydrochlorides, acetates, etc., or bases, are indexed at the formulas of the substances from which they are derived.

The **CAS Number Cross-Reference Index** lists Chemical Abstracts Service Registry Numbers in ascending order together with the corresponding Aldrich Catalog page number.

CD-ROM

The *Sigma-Aldrich Chemical Directory on CD-ROM* contains all the chemicals available from the Sigma-Aldrich Corporation, including the Library of Rare Chemicals. It can be searched by text, physical properties, or structure. The text search uses chemical name or synonym, molecular formula, molecular weight, or Sigma-Aldrich product number. Density, refractive index, boiling point, and melting point searches are available under physical properties. Structure searches can be either imported or drawn on the screen.

INTERNET

The **Aldrich Catalog Handbook of Fine Chemicals** is searchable online at the Aldrich internet web site:

http://www.sigald.sial.com/aldrich/catalog/

The Aldrich online catalog can be searched by text or physical properties. The text search uses chemical name or synonym, CAS registry number, EINECS number, or Sigma-Aldrich product number. Density, refractive index, boiling point, and melting point searches are available in physical properties searches. The matched response leads to a web page containing the chemical name and synonyms, product number, CAS and EINECS numbers, physical properties, references to Aldrich Spectroscopy Libraries and the Beilstein Handbook, and safety data with precautions.

ORGANIZATION

Compounds are arranged alphabetically by systematic name throughout the book. The **Aldrich Catalog/Handbook of Fine Chemicals** is published every two years by the Aldrich Chemical Company, Inc. Individual catalogs highlighting the following specialty areas are also available: Chiral Products, Combinational Chemistry Products, Dyes, Flavors and Fragrances, Fluorinated Products, High Purity Solvents, Indicators and Intermediates, Inorganics and Organometallics, Polymers, Stable Isotopes, Spectroscopy Products, and Rare Chemicals.

EXAMPLE ENTRY

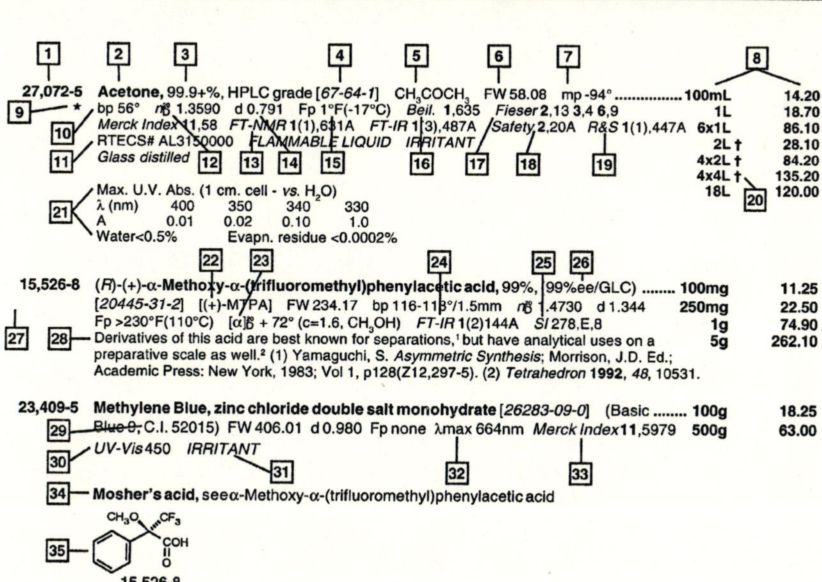

1. Aldrich catalog number
2. Product name
3. Chemical purity
4. Chemical Abstracts Service Registry Number
5. Linear representation of chemical structure
6. Formula weight
7. Melting point
8. Units and prices
9. Denotes that the chemical is in the EPA inventory under TSCA
10. Boiling point
11. Reference to Registry of Toxic Effects of Chemical Substances
12. Index of refraction
13. Reference to *Aldrich Library of ¹³C and ¹H FT NMR Spectra*
14. Density of liquid
15. Flash point
16. Beilstein reference
17. Fieser reference
18. Reference to *Sigma-Aldrich Library of Chemical Safety Data*
19. Reference to *Sigma-Aldrich Library of Regulatory & Safety Data*
20. This quantity of chemical must be shipped via truck
21. Product specifications
22. Alternate product name
23. Specific rotation of a compound determined at the temperature and under the conditions indicated using the D line of sodium
24. Reference to *Aldrich Library of FT-IR Spectra*
25. Reference to *Aldrich Structure Index*, 1996-1997 ed.
26. Percent enantiomeric excess/method of determination
27. Catalog entry has structure appearing at the bottom of page
28. Use statement including pertinent literature reference
29. Colour Index number
30. Reference to *Sigma-Aldrich Library of Stains, Dyes & Indicators*
31. Hazardous properties
32. Wavelength in nanometers at which the maximum absorption of a stain or dye was observed
33. Merck Index reference
34. Cross-reference
35. Structure and catalog number

*Reprinted with permission of Aldrich Chemical Co., Inc., Milwaukee, Wisconsin.

B.2.

Annual Reports in Organic Synthesis:

QD262.A558

- **synthetic conversions**
- **review articles**
- **functional group transformations**

The Reports are annual reviews of synthetically useful information grouped by related reactions and displayed graphically. Emphasis is placed on reactions and methods that are new, general, or useful to synthetic chemists.

EASY ACCESS

— To find a desired transformation or review, check the detailed Table of Contents in the front of each volume and reproduced in the organization section.

INDEXING

The **Author Index** links the name of the senior author, or sometimes the first author, to the corresponding page on which their entry appears.

The Table of Contents serves as the subject index. No cumulative index exists, so each volume must be independently scanned.

ORGANIZATION

Annual Reports in Organic Synthesis is a collection of volumes published every year beginning in 1970. The chapters are as follows:

1. CARBON–CARBON BOND FORMING REACTIONS

    A. Carbon–Carbon Single Bonds (see also: I.E., I.F., I.G., I.H.)
       1. Alkylations of Aldehydes, Ketones, and Their Derivatives
       2. Alkylations of Nitriles, Acids, and Acid Derivatives
       3. Alkylations of β-Dicarbonyl, β-Cyanocarbonyl Systems, and Other Active Methylene Compounds
       4. Alkylations of N-, P-, S-, Se, and Similar Stabilized Carbanions

5. Alkylations of Organometallic and Related Reagents (see also: I.B.3., I.B.4., I.F., I.G.)
6. Other Alkylation Procedures
7. Nucleophilic Addition to Electrophilic Carbon
   a. 1,2-Addition
      (1) Aldol-Type Condensations
      (2) Addition of N-, P-, S-, Se, and Similar Stabilized Carbanions
      (3) Addition of Organometallic and Related Species
      (4) Other 1,2-Additions
   b. Conjugate Additions
      (1) Enolate-Type Carbanions
      (2) Organometallic and Related Reagents
      (3) Other Conjugate Additions
8. Other Carbon–Carbon Single Bond Forming Reactions
B. Carbon–Carbon Double Bonds (See also: I.E.1)
1. Wittig-Type Olefination Reactions
2. Eliminations
   a. Alcohols and Derivatives
   b. Halides
   c. Other Eliminations
3. Other Carbon–Carbon Double Bond Forming Reactions
4. Vinylations
5. Allene Forming Reactions
C. Carbon–Carbon Triple Bonds
D. Cyclopropanations
1. Carbene or Carbenoid Additions to a Multiple Bond
2. Other Cyclopropanations
E. Thermal and Photochemical Reactions
1. Cycloadditions.
2. Other Thermal Reactions
3. Photochemical Reactions
F. Aromatic Substitutions Forming a New Carbon–Carbon Bond
1. Friedel-Crafts Type Aromatic Substitution Reactions
2. Coupling Reactions to Form an Aromatic–Aromatic Bond
3. Other Aromatic Substitutions and Preparations
G. Synthesis via Organometallics
1. Synthesis via Organoboranes
2. Carbonylation Reactions
3. Other Syntheses via Organometallics
H. Rearrangements
1. Claisen, Cope, and Similar Processes
2. Other Rearrangements

II. OXIDATIONS

A. C–O Oxidations
  1. Alcohol → Ketone, Aldehyde
  2. Alcohol, Aldehydes → Acids, Esters
B. C–H Oxidations
  1. C–H → C–O
  2. C–H → C–Hal
C. C–N Oxidations
D. Amine Oxidations
E. Sulfur Oxidations
F. Oxidative Additions to C–C Multiple Bonds
  1. Epoxidations
  2. Hydroxylations
  3. Other Oxidative Additions to C–C Multiple Bonds
G. Phenol–Quinone Oxidation
H. Dehydrogenation
I. Other Oxidations

III. REDUCTIONS

A. C–O Reductions (see also III.F.1)
B. C–N Multiple Bond Reductions
  1. Imine Reductions
  2. Reduction of Heterocycles
C. Reduction of Sulfur Compounds
D. N–O Reductions
E. C–C Multiple Bond Reductions
  1. C=C Reductions
  2. C=C Reductions
F. Hetero Bond Reductions
  1. C–O → C–H
  2. C–Hal → C–H
  3. C–S → C–H
G. Reductive Cleavages
  1. Oxiranes
  2. N–O Cleavage
  3. Other Reductive Cleavages
H. Reduction of Azides

IV. SYNTHESIS OF HETEROCYCLES

A. Oxiranes, Aziridines, and Thiiranes
B. Oxetanes, Azetidines, and Thietanes

C. Lactams
D. Lactones
E. Furans and Thiophenes
F. Pyrroles, Indoles, etc.
G. Pyridines, Quinolines, etc.
H. Pyrans, Pyrones, and Sulfur Analogues
 I. Other Heterocycles with One Heteroatom
 J. Heterocycles with a Bridgehead Heteroatom
K. Heterocycles with Two or More Heteroatoms
  1. Heterocycles with 2 N's
    a. 5-Membered
    b. 6-Membered
    c. 7-Membered
  2. Heterocycles with 2 O's or 2 S's
  3. Heterocycles with 1 N and 1 O
  4. Heterocycles with 1 N and 1 S
  5. Heterocycles with 1 O and 1 S
  6. Heterocycles with 3 or more N's
  7. Heterocycles with 2 N's and 1 O
  8. Heterocycles with 2 N's and 1 S or 1 Se
 L. Other Heterocycles
M. Reviews

V. PROTECTING GROUPS

A. Aldehyde and Ketone Protecting Groups
B. Amino Acid Protection
C. Amine Protecting Groups
D. Carboxyl Protecting Groups
E. Hydroxyl Protecting Groups
F. Other Protecting Groups

VI. USEFUL SYNTHETIC PREPARATIONS

A. Functional Group Preparations
  1. Acetals and Ketals
  2. Acids and Anhydrides (see also: I.G.2.)
  3. Alcohols and Related Species (see also: II.B.1., III.A., V.E., VI.A.9.)
  4. Aldehydes and Ketones (see also: I.A.1., I.G.2., II.A.1)
  5. Amides

6. Amines and Carbamates
7. Amino Acid Derivatives
8. Azides
9. Esters (see also: I.G.2., IV.D., V.D., V.1.A.3)
10. Ethers
11. Halides (see also: II.B.2.)
12. Nitriles and Imines
13. Other N-Containing Functional Groups
B. Additions to Alkenes and Alkynes
C. Nucleotides, etc.
D. Phosphorus, Selenium, and Tellurium Compounds
E. Silicon Compounds
F. Sulfur Compounds
G. Tin Compounds

VII. REVIEWS

A. Techniques
B. Asymmetric Synthesis and Molecular Recognition
C. Reactions
D. Reactive Intermediates
E. Organo-metallics and -metalloids
F. Halogen Compounds and Halogenation (see also: V.1.A.11.)
G. Natural Products
H. Others (see also: IV.M.)

VIII. SELECTED TOPICAL AREAS

A. Fullerene Chemistry
1. Diels-Alder Type Cycloadditions
2. Other Cycloadditions
3. Photochemical Reactions
4. Other Fullerene Chemistry
B. Taxol and Related Taxane Chemistry
C. Enediyne and Dienediyne Chemistry
D. Total Syntheses of Selected Natural Products (see also: VIII.B and VIII.C)
E. Combinatorial Chemistry

AUTHOR INDEX

EXAMPLE ENTRY

I.A.2.   Alkylations of Nitriles, Acids and Acid Derivatives

I.A.2-1 Kusumoto, T. et al., *TL*, **36**, 1071.

62-82%

anti : syn = 10-40:1

*Reprinted from *Annual Reports in Organic Synthesis* (1996), Weintraub PM (ed.) with permission of Academic Press, Inc., Orlando, Florida.

B.3.

Beilstein's Handbook of Organic Chemistry:

QD251.B4921

- compound name and structure
- preparation and derivatives
- physical and chemical properties
- analytical and spectroscopic information

B.3.a. OVERVIEW

The Handbook includes all known data concerning all organic compounds reported in the literature during its period of coverage. This information is researched and critically evaluated before it is included in a compound entry.

EASY ACCESS

— The Centennial Index (blue label on brown binding) contains the General Formula Index and the General Compound Name Index for the entire Handbook.
— General Subject Index is in Volume 28 of any Supplement.
— General Formula Index is in Volume 29 of any Supplement.
— *Aldrich Catalog/Handbook of Fine Chemicals* or the *CRC Handbook of Chemistry and Physics* give the Beilstein reference for their chemicals.
— Use the "Distribution of Compound Types" in B.3.b. to determine the volume in which the compound's functional groups are listed, then look for the compound in the Formula Index or Table of Contents of the indicated volume.

INDEXING

The **Formula Index** lists the molecular formula of chemical substances with their chemical name and Beilstein location. Formulas are arranged by citing carbons in ascending order, then hydrogens, and finally the remaining elements in alphabetical order. If carbon is not present, the elements are arranged in alphabetical order. Stereoisomers that differ only by symbols or prefixes (such as *S* or *trans*) are treated as a single index entry. The Beilstein location is given with the volume number in bold-face type, followed by the corresponding Supplement designation and page number. There are various formula indexes located throughout the Beilstein series.

- The Centennial Index (blue label on brown binding) contains the General Formula Index for all of Beilstein's Handbook.
- Supplement II, Volume 29 is the Cumulative Formula Index for H, EI, and EII.
- Supplements EIII, EIII/IV, EIV, and EV have their own cumulative formula indexes.
- Each volume in supplements EIII, EIII/IV, EIV, and EV has its own individual formula index.

The **Subject Index** or Compound Name Index lists the Beilstein chemical name for a chemical substance and its location in the treatise. For the Original work (H) through Supplement EIII, substituent prefixes are arranged in order of increasing complexity. In Supplement EIII/IV to the present, substituent prefixes are in alphabetical order. For a list of prefixes and their complexity order, see B.3.e. There are various subject indexes located throughout the Beilstein series.

- The Centennial Index (blue label on brown binding) contains the General Compound Name Index for all of Beilstein's Handbook.
- The Cumulative Subject Index for H, EI, and EII is Supplement II, Volume 28.
- The Cumulative Formula Index for H, EI, and EII is Supplement II, Volume 29.
- Supplements EIII, EIII/IV, EIV, and EV have their own cumulative subject indexes.
- Each volume in supplements EIII, EIII/IV, EIV, and EV has its own individual subject index.

COMPUTER SEARCH

The Beilstein database is available by subscribing to the "Crossfire" program for searching the Beilstein in-house database. It is available on STN International by typing FILE BEILSTEIN or through Dialog Information Retrieval Service database [390].

User Information can be assessed by searching the desired Search Field for aids are available by typing HELP or in the STN Database Summary Sheet. For example:

Search Field Name	Code	Search Example	Code
Author	/AU	S CONNERS/AU	
Beilstein Registry No.	/BRN	S 130733/BRN	BRN
CAS Registry Number	/RN	S 89123-39-7/RN	RN
Chemical Name	/CN	S BENZENE/CN	CN
Molecular Formula	/MF	S C3H5/MF	MF
Patent Number	/PN	S 5484983/PN	References
Preparation	/PRE	S ACETONE/PRE	PRE
Reaction	/REA	S PURIDINE/REA	REA

The default display format DISPLAY L(*ist number*) provides information on the Identification of Substance (IDE) plus those display fields in which your search terms appear. The number of occurrences in each field can be accessed with DISPLAY FA. The command DISPLAY ALL gives identification of the substance (IDE), ring system data (RSD), Crossfire reference (XREF), general data (GEN), chemical data (CHE), physical properties (PHY), and all controlled terms (IND). Any of these individual databases can be accessed with the DISPLAY command.

(Copyright 1998 by the American Chemical Society and reprinted with permission.)

INTERNET

Beilstein web page at: http://www.beilstein.com. The site contains various links associated with Beilstein, including database tutorials, dictionaries, and guides.

Beilstein Abstracts is available at http://www.chemweb.com and provides access to title, author, and abstracts of organic journals from 1980 to the present.

STN Database Summary Sheets can be accessed at:

http://info.cas.org/ONLINE/DBSS/dbsslist.html

Dialog Information Retrieval Service Blue Sheets can be accessed at:

http://www.dialog.com/bluesheets/

ORGANIZATION

Beilstein is composed of the following series:

Series (Period)	Abbreviation	Label Color
The Original Work (to 1909)	H	green binding
Supplement I (1910–1919)	EI	red on brown binding
Supplement II (1920–1929)	EII	white binding
Supplement III (1930–1949)	EIII	blue binding
Supplement III/IV (1930–1959)	EIII/IV	blue/black binding
Supplement IV (1950–1959)	EIV	black binding
Supplement V (1960–1979)	EV	red on blue binding

Each Supplement covers the literature for the years indicated and each compound listed is given a System Number based on its Beilstein classification. The Supplements are divided into volumes in the same way as the Original Work, H. At the top center of each page in the Supplements is the corresponding page number as if it were part of the Original Work, H. Therefore, the location of a compound in any one Supplement gives its location in the others.

For cross-referencing, each Supplement gives the page numbers where the same compound can be found in earlier editions of that volume. The System Number for a compound is also indicated at the top of each page and can be used in cross-referencing between supplements.

LITERATURE DESCRIPTION

How to Use Beilstein: Beilstein Handbook of Organic Chemistry, Springer-Verlag: New York, 1978. (QD251.B43 H6)

Online Searching of Beilstein on STN: How to Find Preparations and Reactions; Springer-Verlag: New York, 1990

Sunkel, J., Hoffmann, E., and Luckenbach, R. "Straightforward Procedure for Locating Chemical Compounds in the Beilstein Handbook," *J Chem Ed* 1981; 58:982.

Heller, Stephen R. *The Beilstein System: Strategies for Effective Searching*, American Chemical Society: Washington, D.C., 1998 (QD257.7 B394).

Heller, Stephen R. *The Beilstein Online Database: Implementation, Content, Retrieval*; ACS Symposium Series 436; American Chemical Society: Washington, D.C., 1990 (QD257.7 B39).

EXAMPLE PAGE

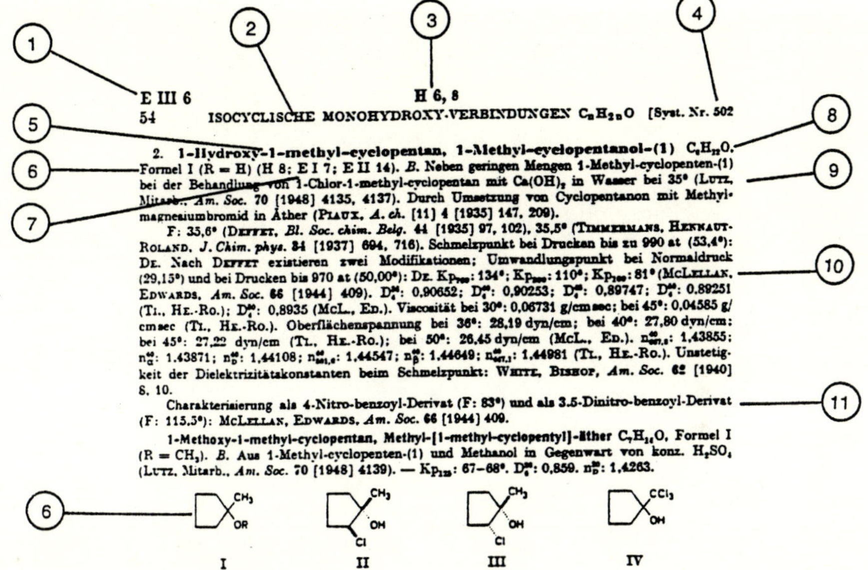

1. Supplement, Volume and Page

2. Section Title and General Formula

3. Corresponding Volume and Page in Orgional Work (H)

4. System Number for Compounds on this Page

5. Important Compound Names

6. Constitutional Formula

7. Back Reference to Previous Supplements

8. Empirical Formula

9. References to Preparation of the Compound

10. References to the Compound's Physical Data

11. References to Derivatives of the Compound

(*Reprinted with permission of Beilstein Institut für Literatur d. Org. Chemie.)

B.3.b.   DISTRIBUTION OF COMPOUND TYPES IN BEILSTEIN

(Compound type followed by volume number.)

A
acetals **1**, cyclic **19**
acetylenes **1**, alicyclic **5**
acid chlorides **2**, alicyclic **9**
acid halides **2**, alicyclic **9**
acids **2**, alicyclic **9**
acridine **20**, derivatives **21, 22**
alcohols **1**, alicyclic **6**
aldehydes **1**, alicyclic **7**
alkenes **1**, alicyclic **5**
alkynes **1**, alicyclic **5**
amides **2**, alicyclic **17**
amines **4**, alicyclic **12**, aromatic **20**, cyclic **20**
amino acids **4**, alicyclic **14**
anhydrides **2**, alicyclic **9**, cyclic **17**
aromatics **5**
azocarboxylate **3**, alicyclic **10**

B
benzimidazole **23**, derivatives **24, 25**
benzofuran **17**, derivatives **18**
benzothiophene **17**, derivatives **18**
bromides **1**, alicyclic **5**

C
carbazole **20**, derivatives **21, 22**
carbodiimides **4**, alicyclic **12**
carbonates **1**, alicyclic **6**
carbonates **3**, alicyclic **10**, cyclic **19**
carboxylic acids **2**, alicyclic **9**
chlorides **1**, alicyclic **5**
chloroformates **1**, alicyclic **6**
cinnoline **23**, derivatives **24, 25**
cyanohydrins **3**, alicyclic **10**

D
dioxane **19**
dioxolane **19**
dithiane **19**

E
epoxides **17**
esters **2**, alicyclic **9**
ethers **1**, alicyclic **6**, cyclic **17**

F
fluorides **1**, alicyclic **5**
furan **17**, derivatives **18**

G
guanidines **4**, alicyclic **12**

H
halides **1**, alicyclic **5**
hydrazines **4**, alicyclic **15**
hydroperoxides **1**
hydroxylamines **4**, alicyclic **15**

I
imidazole **23**, derivatives **24**, **25**
imides **2**, alicyclic **9**, cyclic **21**
indazole **23**, derivatives **24**, **25**
indole **20**, derivatives **21**, **22**
indolizine **20**, derivatives **21**, **22**
iodides **1**, alicyclic **5**
isocyanates **4**, alicyclic **12**
isoquinoline **20**, derivatives **21**, **22**
isothiocyanate **4**, alicyclic **12**
isoxazole **27**

K
ketones **1**, alicyclic **7**

L
lactones **17**

M
monosaccharides **31**
morpholine **27**

N

nitriles **2**, alicyclic **9**
nitro **1**, alicyclic **5**
nucleic acids **1**
nucleotides **31**

O

olefins **1**, alicyclic **5**
oxazole **27**
oxazolidone **27**
oximes **1**, alicyclic **7**

P

peracids **2**, alicyclic **9**
phenazine **23**, derivatives **24, 25**
phenols **6**
phenothiazine **27**
phenoxazine **27**
phosphines **4**, alicyclic **16**
phosphite **1**
phosphonic acid **4**, alicyclic **16**
phthalazine **23**, derivatives **24, 25**
piperazine **23**, derivatives **24, 25**
piperidine **20**, derivatives **21, 22**
pteridine **26**
purine **26**
pyrazine **23**, derivatives **24, 25**
pyrazole **23**, derivatives **24, 25**
pyridazine **23**, derivatives **24, 25**
pyridine **20**, derivatives **21, 22**
pyrimidine **23**, derivatives **24, 25**
pyrrole **20**, derivatives **21, 22**
pyrrolidine **20**, derivatives **21, 22**
pyrroline **20**, derivatives **21, 22**

Q

quinazoline **23**, derivatives **24, 25**
quinoline **20**, derivatives **21, 22**
quinuclidine **20**, derivatives **21, 22**

S

sugars **31**
sulfides **1**
sulfone **1**, alicyclic **17**
sulfonamides **4**, alicyclic **11**
sulfonate esters **4**, alicyclic **11**
sulfonates **4**, alicyclic **11**
sulfones **1**, alicyclic **17**
sulfonic acids **4**, alicyclic **11**
sulfonyl chlorides **4**, alicyclic **11**
sulfoxides **1**, alicyclic **17**

T

thiazole **27**
thiols **1**, alicyclic **6**
thiomorpholine **27**
thiophene **17**, derivatives **18**
triazole **26**
trithiane **19**

U

ureas **4**, alicyclic **12**

B.3.c.   ABBREVIATIONS

(*Reprinted with permission of Beilstein Institut für Literatur d. Org. Chemie.)

A complete list of journal abbreviations is located at the beginning of each volume.

Note: The use of capital letters and/or a period following the abbreviation is significant. Journal abbreviations are italicized.

A
A. = Liebigs Annalen der Chemie
A. = ethanol
A. ch. = Annales de Chimie
Acn. = acetone
Ae. = diethyl ether
akt. = active
alkal. = alkaline
alkoh. = alcoholic
Am. Soc. = Journal of the American Chemical Society
ang. = angular
Ang. Ch. = Angewandte Chemie
Anm. = annotation, footnote
A.P. = American patent
App. = apparatus
ar = aromatic
as = asymmetric
asymm. = asymmetric
at = technical atmosphere
äther. = ethereal, essential, volatile
atm = physical (standard) atmosphere
Aufl. = edition

B
B. = Berichte der Deutschen Chemischen Gesellschaft
B. = formation, synthesis
Bd. = volume
ber. = calculated
Bildg. = formation
Bl. = Bulletin de la Societe Chimique de France
Bzl. = benzene
Bzn. = petroleum ether
bzw. = or, respectively

C

C = concentration
C. = Chemisches Zentralblatt
C. A. = Chemical Abstracts
C. r. = Comptes Rendus Hebdomadaires de Seances
de l'Academie des Sciences

D

d = day
D. = (1) Debye; (2) density, specific gravity
Darst. = preparation, representation
D.B.P. = German patent (Deutsches Bundespatent)
Diss. = dissertation
D.P. = German patent
D.R.P. = German patent (Deutsches Reichspatent)
[D(R − X)] = dissociation energy

E

E = (1) freezing (solidification) point; (2) Beilstein
Supplementary Series
E. = ethyl acetate
Eg. = acetic acid
Eig. = property
Einw. = action, influence
E.K. = E.M.F.
E.M.K. = E.M.F.
engl. Ausg. = English edition
engl. Ausgabe = English edition
E.P. = English patent
EPR = electron paramagnetic resonance (= ESR)
Ergw. = Beilstein Supplementary Series

F

F = melting point (range)
F.P. = French patent

G

G. = Gazzetta Chimica Italiana
gem. = geminal
Gew.-% = percent by weight

Ggw. = presence
grad = Grad = degree

H
h = hour(s)
H = Beilstein Basic Series
Helv. = Helvetica Chimica Acta
Hptw. = Beilstein Basic Series

I
inakt. = inactive

J
J.pr. = Journal für Praktische Chemie

K
konz. = concentrated
Konz. = concentration
korr. = corrected
Kp = boiling point

L
l = (1) liter; (2) length of tube in dm (polarimetry)
l.c. = loco citato (in the place cited)
Leg. = alloy
lin. = linear

M
m = (1) meter; (2) molarity of solution
M. = Monatshefte für Chemie
Me. = methanol
Meth. = method (of)
min = minute
Mitarb. = coworker, collaborator
Mol. Gew. = molecular weight
ms- = meso

N
n = (1) number of formula units in a unit cell;
(2) normality of solution; (3) nano (= 10^{-9});

(4) refractive index
Nd. = precipitate

O
O.P. = Austrian patent
opt.-akt. = optically active
opt.-inakt. = optically inactive

P
PAe. = petroleum ether
P.C.H. = Pharmazeutische Zentralhalle für Deutschland
prim. = primary
Priv.-Mitt. = private communication
Py. = pyridine

R
R. = Recueil des Travaux Chimiques des Pays-Bas
rac. = racemic
R.I.-Bezfg. = ring index numbering
RIS = Ring Index Supplement
Rk. = reaction
RRI = Ring Index (2nd ed.; 1960)
RV = reducing power

S
s = second
S. = page
s. = see
s.a. = see also
Sci. = Science, New York
sek. = secondary
snF = see adjoining formula
s.o. = see above
Soc. = Journal of the Chemical Society, London
sog. = so-called
spezif. = specific
Spl. = supplement (of a book)
...stdg. = for ... hours
s.u. = see below

Syst.Nr. = System Number in Beilstein
symm. = symmetric

T
Tl. = part

U
U.S.P. = United States patent
unverd. = undiluted
unkorr. = uncorrected

V
V = occurrence
verd. = diluted
vgl. = compare
vic. = vicinal
Vol. = volume
Vork. = occurrence

W
W. = water
wässr. = water, aqueous
wss. = aqueous

Z
z.B. = for example
Zers. = decomposition
zit.bei = cited in
Z.*Kr*. = Zeitschrift für Kristallographie,
        Kristallgeometrie, Kristallphysik,
        Kristallchemie

B.3.d.   RULES OF THE BEILSTEIN SYSTEM OF INDEXING

1. The organic compounds are distributed over the Beilstein volumes as follows:

Table 4: Contents of the 27 Volumes of the Beilstein Handbook

Type of registry compound	Feature of the functional group	A (Acyclics)	B (Isocyclics)	C (Heterocyclics) — Type and number of ring heteroatoms					
				1O*)	2O*), 3O*),....	1N	2N	3N, 4N,...	1N, 1O*), 1N, 2O*),.... 2N, 1O*), 2N, 2O*),.... further hetero-atoms**)
(1) Compounds without functional groups	–		5			20			
(2) Hydroxy-compounds	$-OH$	1	6	17			23		
(3) Oxo-compounds	$=O$		7			21	24		
	$=O + -OH$		8						
(4) Carboxylic acids $<^{=O}_{OH}$	$<^{=O}_{OH}$; $\left[<^{=O}_{OH}\right]_n$	2	9						
	$<^{=O}_{OH} + -OH$; $<^{=O}_{OH} + =O$; $<^{=O}_{OH} + =O + -OH$	3	10						
(5) Sulfinic acids	$-SO_2H$								
(6) Sulfonic acids	$-SO_3H$								
(7) Seleninic acids, Selenonic acids, Tellurinic acids	$-SeO_2H$ and $-SeO_3H$ $-TeO_2H$	4	11	18	19	22	25	26	27
(8) Amines $-NH_2$	$-NH_2$		12						
	$[-NH_2]_n$; $-NH_2 + -OH$		13						
	$-NH_2 + =O$; $-NH_2 + <^{=O}_{OH}$; $-NH_2 + ...$		14						
(9) Hydroxylamines and Dihydroxyamines	$-NH-OH$ $-N<^{OH}_{OH}$		15						
(10) Hydrazines	$-NH-NH_2$								
(11) Azo-compounds	$-N=NH$								
(12) Diazonium compounds	$-N\equiv N|^{\oplus}$								
(13) Compounds with groups of 3 or more N-atoms	$-NH-NH-NH_2$, $-N(NH_2)_2$, $-N=N-NH_2$, etc.								
(14) Compounds containing carbon directly bonded to P, As, Sb, and Bi	e.g. $-PH_2$, $PH-OH$, $-P(OH)_2$, $-PH_4$,...., $-PO(OH)_2$		16						
(15) Compounds containing carbon directly bonded to Si, Ge, and Sn	e.g. $-SiH_3$, $-SiH_2(OH)$, ...								
(16) Compounds containing carbon directly bonded to elements of the 3rd – 1st A-groups of the periodic table	e.g. $-BH_2$, $-BH(OH)$, ..., $-Mg^{\oplus}$								
(17) Compounds containing carbon directly bonded to elements of the 1st – 8th B-groups of the periodic table	e.g. $-HgH$, $-Hg^{\oplus}$,....								

*) instead of O also S, Se, Te (cf. p. 25 f.)   **) e. g. B, Si, P, but not S, Se, Te (cf. p. 25 f.)

2. Compounds within the classes given in (1) are further ordered by the type and number of functional groups.

(a) Compounds with nonfunctional substituents (F, Cl, Br, I, NO, NO_2, N_3) are found in the same volume as the compounds *without* these substituents.

$Cl_3C—CHO$ $H_3C—CHO$

Vol.1   Vol.1   Vol. 5    Vol. 5

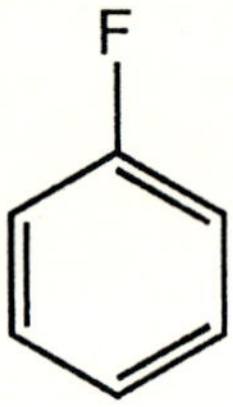

(b) Compounds with several *identical* functional groups follow those that have only *one* such functional group.

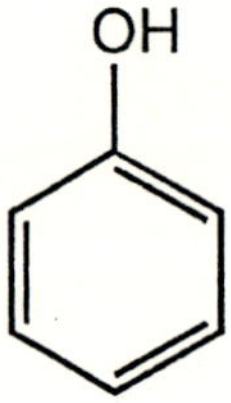

Vol. 6 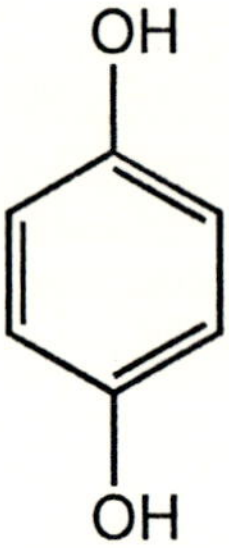   Vol. 6 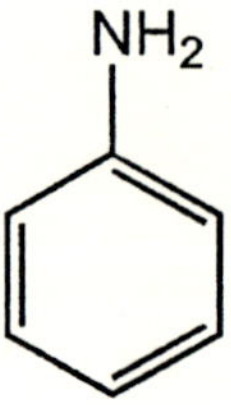   Vol. 12 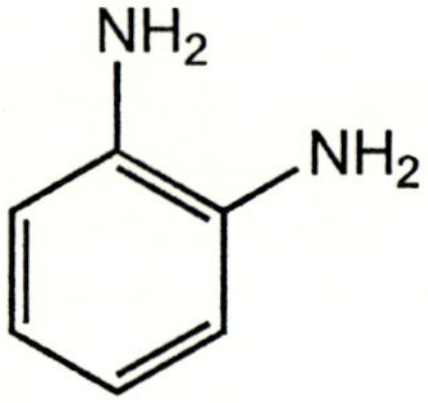   Vol. 13

(c) Compounds with several *different* functional groups are described in Beilstein at the *"latest possible systematic entry,"* i.e., the functional group with the highest (encircled) number in (1) is the factor determining its allocation.

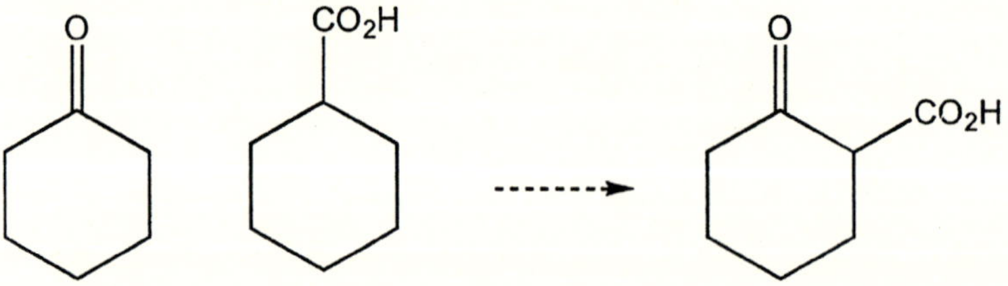

Vol. 7   Vol. 9   Vol. 10

3. Compounds containing one or more *derivatized* functional groups (e.g., —COCl) are considered as *functional derivatives*. These are subjected to formal hydrolysis and are so reconverted to compounds with free functional groups (e.g., —COOH), which can be located in Beilstein from (1). The required *derivative* is then described following the systematically latest compound with free functional groups. Where various routes are available for formal hydrolysis, that route is chosen which leads to the compound described at the *latest possible* place in Beilstein (i.e., in the volume with the *highest possible* volume number).

e.g.

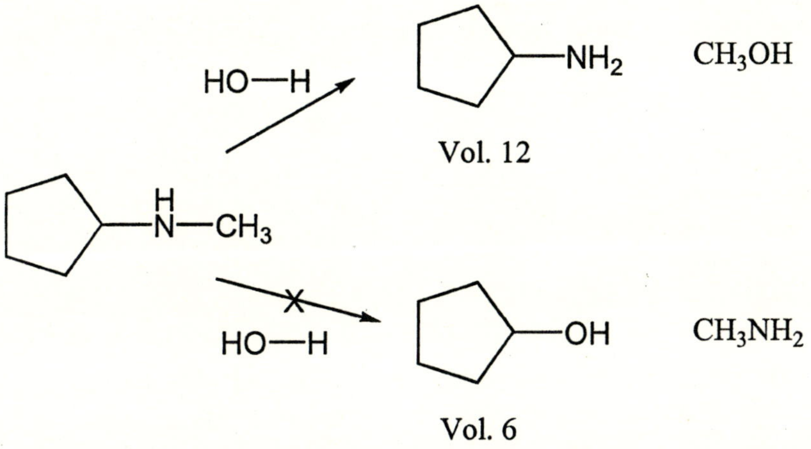

$$HO\text{—}H \longrightarrow \text{(cyclopentyl)}\text{—}NH_2 \qquad CH_3OH$$

Vol. 12

$$\text{(cyclopentyl)}\text{—}\overset{H}{N}\text{—}CH_3$$

$$HO\text{—}H \xrightarrow{\;\;\times\;\;} \text{(cyclopentyl)}\text{—}OH \qquad CH_3NH_2$$

Vol. 6

Compounds carrying a functional *and* a nonfunctional group on *one* carbon atom are also subjected to formal hydrolysis.

4. Organic S, Se, and Te compounds are considered in Beilstein as *formal derivatives* of the corresponding oxygen compounds and are described following these. Sulfoxides, sulfones, and sulfonium compounds are dealt with after the corresponding sulfides.

LITERATURE DESCRIPTION

Sunkel, J., Hoffmann, E., Luckenbach, R. "Straightforward Procedure for Locating Chemical Compounds in the Beilstein Handbook," *J. Chem Ed* 1981; *58*:982.

B.3.e.   LIST OF PREFIXES IN BEILSTEIN'S HANDBOOK

A complete list of Beilstein's prefixes is located at the beginning of 17/1 of EIII/IV (blue pages), in all the volumes in EV (yellow pages), and at the beginning of the Centennial Index — General Compound Index (yellow pages).

(*Reprinted with permission of Beilstein Institut für Literatur d. Org. Chemie.)

Formula	Structure	Prefix	Priority
Al^{2+}	$-Al^{2+}$	Aluminio$^{(2+)}$	
As	$-As<$	Arsantriyl	
AsClHO	$-ASH(O)-Cl$	Chloroarsinoyl	
$AsHO_2$	$>As(O)-OH$	Hydroxyarsoryl	
AsH_2	$-AsH_2$	Arsino	162
AsH_2O	$-AsH_2O$	Arsinoyl	
AsH_2O_3	$-AsO(OH)_2$	Arsono	168
AsH_3	$=AsH_3$	Arsoranyliden	
AsH_3^+	$-AsH_3^+$	Arsonio	
AsH_4	$-AsH_4$	Arsoranyl	
AsO	$-AsO$	Arsenoso	164
	$-As(O)<$	Arsoryl	
AsO_2	$-AsO_2$	Arso	167
As_2H_3	$-AsH-AsH_2$	Diarsenyl	
B	$-B<$	Borantriyl	
BH_2	$-BH_2$	Bornyl	182
B_2H_5	$-B_2H_5$	Diboran(6)-yl	
Bi^{2+}	$-Bi^{2+}$	Bismutio	
BiH_2	$-BiH_2$	Bismutino	173

Formula	Group	Name	Page
Br	$-Br$	Brom	3
Br^-	Br^-	Bromo	
C	$>C<$	Methantetrayl	
CClO	$-CO-Cl$	Chlorformyl	331
CH	$>CH-$	Methantriyl	
	$=CH-$	Methanylyliden	
	$=CH-$	Metheno	
	$\equiv CH$	Methylidine	
CHN	$-C(=NH)-$	Carbimidoyl	
CHO	$-CHO$	Formyl	276
CHOS	$-CO-SH$	Thiocarboxy	337
CHO_2	$-CO-OH$	Carboxy	326
CHO_2S	$-S-CO-OH$	Carboxymercapto	103
CHO_3	$-O-CO-OH$	Carboxyoxy	88
CHS	$-CHS$	Thioformyl	278
CHS_2	$-CS-SH$	Dithiocarboxy	340
CH_2	$-CH_2-$	Methandiyl	
	$-CH_2-$	Methano	
	$=CH_2$	Methylen	255
CH_2Cl	$-CH_2-Cl$	Chlormethyl	188
CH_2N	$-CH=NH$	Formimdoyl	277
	$-CH=NH$	Iminomethyl	
CH_2NO	$-CO-NH_2$	Carbamoyl	333
	$-CH=N-OH$	Formohydroximoyl	
CH_2NO_2	$-NH-CO-OH$	Carboxyamino	41
CH_2NS	$-CS-NH_2$	Thiocarbamoyl	339

Formula	Structure	Prefix	Priority
CH_2N_2O	$-NH-CO-NH-$	Ureylen	44
CH_2N_2S	$-NH-CS-NH-$	Thioureylen	47
CH_2O	$-O-CH_2-$	Oxaäthano	
CH_2O_2	$-O-CH_2-O-$	[1,3]Dioxapropano	
	$-O-CH_2-O-$	Methylendioxy	82
CH_3	$-CH_3$	Methyl	187
CH_3ClN	$-NH-CH_2Cl$	Chlormethyl-amino	
	$-N(Cl)-CH_3$	Chlormethyl-amino	
CH_3N	$-NH-CH_2-$	Azaäthano	
CH_3N_2	$-C(NH_2)=NH$	Carbamimidoyl	
	$-CH=N-NH_2$	Formohydrazonoyl	
CH_3N_2O	$-CO-NH-NH_2$	Carbazoyl	
	$-NH-C(OH)=NH$	Isoureido	50
	$-NH-CO-NH_2$	Ureido	42
CH_3N_2S	$-NH-C(SH)=NH$	Isothioureido	48
	$-NH-CS-NH_2$	Thioureido	47
CH_3N_3O	$=N-NH-CO-NH_2$	Semicarbazono	127
CH_3N_2S	$=N-NH-CS-NH_2$	Thiosemicarbazono	147
CH_3N_4	$-NH-NH-CO-N=NH$	Formazano	
	$-(N=NH)=N-NH_2$	Formazanyl	
CH_3N_4O	$-NH-NH-CO-N=NH$	Carbazono	
CH_3O	$-CH_2-OH$	Hydroxymethyl	190
	$-O-CH_3$	Methoxy	67
CH_3OS	$-SO-CH_3$	Methyansulfinyl	105
CH_3O_2S	$-SO_2-CH_3$	Methansulfonyl	107
CH_3S	$-S-CH_3$	Methansulfenyl	
	$-S-CH_3$	Methylmercapto	97

Formula	Structure	Name	
CH_3S^-	CH_3-S^-	Methanthiolato	
CH_3Se	$-Se-CH_3$	Methylselanyl	
CH_3Te	$-Te-CH_3$	Methyltellanyl	
CH_4N	$-CH_2-NH_2$	Aminomethyl	189
CH_4N_2	$-NH-C(NH_2)=NH$	Guanidino	45
CH_4N_3O	$-NH-NH-CO-NH_2$	Semicarbazido	146
CH_4NN_4O	$-NH-NH-CO-NH-NH_2$	Carbonohydrazido	
CN	$-CN$	Cyan	336
	$-NC$	Isocyan	34
CN^-	CN^-	Cyano	
CNO	$-OCN$	Cyanato	90
	$-NCO$	Isocyanato	49
CNS	$-NCS$	Isothiocyanato	52
	$-SCN$	Thiocyanato	104
$CNSe$	$-NCSe$	Isoselenocyanato	
	$-SeCN$	Selenocyanato	113
CO	$-CO-$	Carbonyl	
CO_3	$-O-CO-O-$	Carbonyldioxy	89
CO_3^{2-}	CO_3^{2-}	Carbonato	
CS	$-CS-$	Thiocarbonyl	
C_2	$-C{\equiv}C-$	Äthinyl	
C_2ClO_2	$-CO-CO-Cl$	Chloroxalyl	
C_2H	$-C{\equiv}CH$	Äthinyl	253
C_2HO_2	$-CO-CHO$	Glyoxyloyl	
C_2HO_2S	$-SO_2-C{\equiv}CH$	Äthinsulfonyl	
C_2HO_3	$-CO-CO-OH$	Hydroxyoxalyl	

Formula	Structure	Prefix	Priority
C_2H_2	$=CH-CH=$	Äthandiyliden	
	$-CH=CH-$	Äthendiyl	
	$-CH=CH-$	Ätheno	
	$=C=CH_2$	Vinyliden	
$C_2H_2NO_2$	$-CO-CO-NH_2$	Aminooxalyl	
$C_2H_2N_2$	$-C(=NH)-C(=NH)-$	Oxalimidoyl	
C_2H_3	$-CH_2CH<$	Äthantriyl	
	$-CH_2-CH=$	Äthylyliden	
	$\equiv C-CH_3$	Äthylidin	
	$-CH=CH_2$	Vinyl	207
$C_2H_3HgO_2$	$-Hg-O-CO-CH_3$	Acetatomercurio	186
C_2H_3NO	$=N-CO-CH_3$	Acetylimino	123
$C_2H_3N_2O_2$	$-CO-NH-CO-NH_2$	Allophanoyl	
C_2H_3O	$-CO-CH_3$	Acetyl	279
C_2H_3OS	$-S-CO-CH_3$	Acetylmercapto	102
$C_2H_3O_2$	$-O-CO-CH_3$	Acetoxy	85
	$-CH_2-CO-OH$	Carboxymethyl	192
	$-CO-CH_2-OH$	Glykoloyl	
$C_2H_3O_2^-$	CH_3-CO-O^-	Acetato	
$C_2H_3O_2S$	$-SO_2-CH=CH_2$	Äthensulfonyl	
$C_2H_3O_3$	$-O-O-CO-CH_3$	Acetylperoxy	
C_2H_3S	$-CS-CH_3$	Thioacetyl	281
$C_2H_3S_2$	$-S-CS-CH_3$	Thioacetylmercapto	
C_2H_4	$-CH_2-CH_2-$	Äthylen	
	$-CH_2-CH_2-$	Äthano	249

Formula	Group	Name	Page
	$=CH-CH_3$	Äthyliden	256
	$> CH-CH_3$	Äthyliden	256
C_2H_4N	$-C(CH_3)=NH$	Acetimidoyl	
	$-C(CH_3)=NH$	1-Imino-äthyl	
C_2H_4NO	$-C(CH_3)=N-OH$	Acetohydroxyimoyl	
C_2H_4NO	$-NH-CO-CH_3$	Acetylamino	
	$-CO-CH_2-NH_2$	Glycyl	355
	$-C(CH_3)=N-OH$	1-Hydroxyimino-äthyl	
$C_2H_4N_4O_2$	$-NH-CO-NH-NH-CO-NH-$	Biureylen	
C_2H_4O	$-CH_2-O-CH_2-$	Oxapropano	
$C_2H_4O_2$	$-O-CH_2-CH_2-O-$	Äthylendioxy	81
	$-O-CH(CH_3)-O-$	Äthylidenedioxy	
C_2H_5	$-C_2H_5$	Äthyl	194
C_2H_5N	$=N-C_2H_5$	Äthylimino	
	$-NH-CH_2-CH_2-$	[1]Azapropano	
	$-CH_2-NH-CH_2-$	[2]Azapropano	
$C_2H_5N_2$	$-C(CH_3)=N-NH_2$	Acetohydrazonoyl	
	$-N=N-C_2H_5$	Äthylazo	
	$-C(CH_3)=N-NH_2$	1-Hydrazono-äthyl	
C_2H_5O	$-O-C_2H_5$	Äthoxy	68
$C_2H_5O^-$	$C_2H_5O^-$	Äthoxo	
C_2H_5OS	$-SO-C_2H_5$	Äthansulfinyl	
C_2H_5Se	$-SeO-C_2H_5$	Äthanseleninyl	
$C_2H_5O_2$	$-O-O-C_2H_5$	Äthylperoxy	
$C_2H_5O_2S$	$-SO_2-C_2H_5$	Äthansulfonyl	
$C_2H_5O_2Se$	$-SeO_2-C_2H_5$	Äthanselenonyl	

Formula	Structure	Prefix	Priority
C_2H_5S	$-S-C_2H_5$	Äthansulfenyl	
	$-S-C_2H_5$	Äthylmercapto	98
C_2H_5S-	C_2H_5S-	Äthanthiolato	
$C_2H_5S_2$	$-S-S-C_2H_5$	Äthyldisulfanyl	
C_2H_5Se	$-Se-C_2H_5$	Äthanselenenyl	
C_2H_5Se	$-Se-C_2H_5$	Äthylselanyl	
C_2H_5Te	$-Te-C_2H_5$	Äthyltellanyl	
$C_2H_6AsO_2$	$-AsH(O)-OC_2H_5$	Äthoxyarsinoyl	
C_2H_6N	$-NH-C_2H_5$	Äthylamino	26
$C_2H_6O_2P$	$-PH(O)-OC_2H_5$	Äthoxyphosphinoyl	
C_2O_2	$-CO-CO-$	Oxalyl	309
C_3HO	$-CO-C{\equiv}CH$	Propioloyl	298
$C_3H_2O_2$	$CO-CH_2-CO-$	Malonyl	310
C_3H_3	$-C{\equiv}C-CH_3$	Propinyl	216
C_3H_3O	$-CO-CH{=}CH_2$	Acryloyl	293
$C_3H_3O_2$	$-CO-CO-CH_3$	Pyruvoyl	350
$C_3H_3O_3$	$-CO-CH_2-CO-OH$	Carboxyacetyl	
C_3H_4	$={CH}-CH{=}CH_2$	Allyliden	263
	$-CH{=}CH-CH_2-$	Propeno	
$C_3H_4NO_2$	$-CO-CH_2-CO-NH_2$	Malonamoyl	
C_3H_4O	$={CH}-CO-CH_3$	Acetonyliden	274
C_3H_5	$-CH_2-CH{=}CH_2$	Allyl	209
	$-C(CH_3){=}CH_2$	Isopropenyl	210
	$-CH{=}CH-CH_3$	Propenyl	208

Formula	Structure	Name	Page
$C_3H_5N_2O_3$	$-CO-CH_2-NH-CO-NH_2$	Hydantoyl	
C_3H_5O	$-CH_2-CO-CH_3$	Acetonyl	269
	$-CO-CH_2-CH_3$	Propionyl	282
C_3H_5OS	$-CO-S-C_2H_5$	Äthylmercaptocarbonyl	
$C_3H_5O_2$	$-CO-OC_2H_5$	Äthoxycarbonyl	328
	$-CO-CH(OH)-CH_3$	Lactoyl	342
$C_3H_5O_3$	$-CO-CH(OH)-CH_2-OH$	Glyceroyl	
C_3H_6	$=C(CH_3)_2$	Isopropyliden	258
	$-CH_2-CH_2-CH_2-$	Propandiyl	
	$-CH_2-CH_2-CH_2-$	Propano	
	$=CH-CH_2-CH_3$	Propyliden	257
C_3H_6NO	$-C(OC_2H_5)=NH$	Äthoxycarbimidoyl	
	$-CO-CH(NH_2)-CH_3$	Alanyl	358
	$-CO-CH_2-CH_2-NH_2$	β-Alanyl	359
	$-CO-CH_2-NH-CH_3$	N-Methyl-glycyl	
	$-NH-CO-CH_2-CH_3$	Propionylamino	38
	$-CO-CH_2-NH-CH_3$	Sarkosyl	356
C_3H_6NOS	$-CO-CH(NH_2)-CH_2-SH$	Cysteinyl	378
$C_3H_6NO_2$	$-C(OC_2H_5)=NOH$	Äthoxycarbohydroximoyl	
	$-NH-CO-OC_2H_5$	Äthoxycarbonylamino	
	$-NH-CH(CH_3)-CO-OH$	Alanin-N-yl	
	$-CO-CH(NH_2)-CH_2-OH$	Seryl	377
$C_3H_6NO_2S$	$-S-CH_2-CH(NH_2)-CO-OH$	Cystein-S-yl	
$C_3H_6NO_4S$	$-CO-CH(NH_2)-CH_2-SO_2-OH$	Cysteyl	
C_3H_7	$-CH(CH_3)_2$	Isopropyl	196
	$-CH_2-CH_2-CH_3$	Propyl	195

Formula	Structure	Prefix	Priority
C_3H_7O	$-O-CH(CH_3)_2$	Isopropoxy	70
	$-O-CH_2-CH_2-CH_3$	Propoxy	69
$C_3H_7O_2S$	$-SO_2-CH(CH_3)_2$	Propansulfonyl	
C_3H_8N	$-CH_2-N(CH_3)_2$	Dimethylaminomethyl	
C_3O_3	$-CO-CO-CO-$	Mesoxalyl	353
$C_4H_2O_2$	$-CO-CH=CH-CO-$	Butendioyl	
	$-CO-CH=CH-CO-$ (*trans*)	Fumaroyl	320
	$-CO-CH=CH-CO-$ (*cis*)	Maleoyl	319
$C_4H_2O_3$	$-CO-CO-CH_2-CO-$	Oxalacetyl	
$C_4H_4O_2$	$-CO-CH(OH)-CH_2-CO$	Maloyl	354
	$-CO-CH_2-CH_2-CO-$	Succinyl	312
$C_4H_4O_4$	$-CO-CH(OH)-CH(OH)-CO-$ (*threo*)	Tartaroyl	
$C_4H_5NO_2$	$-CO-CH(NH_2)-CH_2-CO-$	Aspartoyl	372
C_4H_5O	$-CO-CH_2-CH=CH_2$	But-3-enoyl	
	$-CO-CH=CH-CH_3$	Crotonoyl	294
	$-CO-C(CH_3)=CH_2$	Methacryloyl	296
$C_4H_5O_2$	$-CO-CH_2-CO-CH_3$	Acetoacetyl	351
$C_4H_5O_3$	$-O-CO-CH_2-CO-CH_3$	Acetoacetyloxy	
	$-CO-CO-0C_2H_5$	Äthoxyoxalyl	307
C_4H_6	$-CH=CH-CH_2-CH_2-$	But-1-eno	
	$-CH_2-CH=CH-CH_2-$	But-2-eno	
$C_4H_6IO_4$	$-I(O-CO-CH_3)_2$	Diacetoxyjod	8
$C_4H_6NO_2$	$-N(CO-CH_3)_2$	Diacetylamino	37
	$-CO-CH_2-CH_2-CO-NH_2$	Succinamoyl	311

Formula	Structure	Name	Page
$C_4H_6NO_3$	$-CO-CH(NH_2)-CH2-COOH$	α-Aspartyl	369
	$-CO-CH_2-CH(NH_2)-COOH$	β-Aspartyl	371
C_4H_7	$-CH_2-CH=CH-CH_3$	Butenyl	212
	$-CH_2-C(CH_3)=CH_2$	Methallyl	214
$C_4H_7N_2O_2$	$-CO-CH(NH_2)-CH_2-CO-NH_2$	Asparaginyl	370
	$-CO-CH_2-CH(NH_2)-CO-NH_2$	Isoasparaginyl	
C_4H_7O	$-CO-CH_2-CH_2-CH_3$	Butyryl	283
	$-CO-CH(CH_3)_2$	Isobutyryl	284
C_4H_7S	$CS-CH(CH_3)_2$	Thioisobutyryl	
C_4H_8	$-CH_2-CH_2-CH_2-CH_2-$	Butandiyl	
C_4H_8	$-CH_2-CH_2-CH_2-CH_2-$	Butano	
	$=CH-CH_2-CH_2-CH_3$	Butyliden	
C_4H_8NOS	$-CO-CH(NH_2)-CH_2-CH_2-SH$	Homocysteinyl	
$C_4H_8NO_2$	$-CO-CH(NH_2)-CH(OH)-CH_3 (erythro)$	Allothreonyl	
	$-CO-CH(NH_2)-CH_2CH_2OH$	Homoseryl	
	$-CO-CH(NH_2)-CH(OH)-CH_3 (threo)$	Threonyl	380
C_4H_9	$-CH_2-CH_2-CH_2-CH_3$	Butyl	197
	$-CH(CH_3)-CH_2-CH_3$	sec-Butyl	198
	$-C(CH_3)_3$	tert-Butyl	200
	$-CH_2-CH(CH_3)_2$	Isobutyl	199
C_4H_9O	$-O-CH_2-CH_2-CH_2-CH_3$	Butoxy	71
	$-O-CH(CH_3)-CH_2-CH_3$	sec-Butoxy	72
	$-O-C(CH_3)_3$	tert-Butoxy	74
	$-O-CH_2-CH(CH_3)_2$	Isobutoxy	73
$C_4H_{10}AsO_3$	$-AsO(OC_2H_5)_2$	Diäthoxyarsoryl	
$C_4H_{10}N$	$-N(C_2H_5)_2$	Diäthylamino	
$C_4H_{10}OP$	$-PO(C_2H_5)_2$	Diäthylphosphinoyl	

Formula	Structure	Prefix	Priority
$C_4H_{10}O_2PS$	$-PS(OC_2H_5)_2$	Diäthoxythiophosphoryl	
$C_4H_{10}O_3P$	$-PO(OC_2H_5)_2$	Diäthoxyphosphoryl	
$C_5H_7O_2$	$-CO-CH_2-CH_2-CH_2-CO-$	Glutamyl	313
$C_5H_7NO_2$	$-CO-CH(NH_2)-CH_2-CH_2-CO-$	Glutamoyl	376
C_5H_7O	$-CO-C(CH_3)=CH-CH_3(Z)$	Angeloyl	
	$-CO-C(CH_3)=CH-CH_3(E)$	Tigloyl	
$C_5H_7O_2$	$-CO-CH_2-CH_2-CO-CH_3$	Lävulinoyl	
$C_5H_8NO_3$	$-CO-CH(NH_2)-[CH_2]_2-CO-OH$	α-Glutamyl	373
$C_5H_8NO_3$	$-CO-[CH_2]_2--CH(NH_2)-CO-OH$	γ-Glutamyl	375
$C_5H_9N_2O_2$	$-CO-CH(NH_2)-[CH_2]_2-CO-NH_2$	Glutaminyl	374
	$-CO-[CH_2]_2-CH(NH_2)-CO-NH_2$	Isoglutaminyl	
C_5H_9O	$-CO-CH_2-CH(CH_3)_2$	Isovaleryl	286
	$-CO-C(CH_3)_3$	Pivaloyl	287
	$-CO-[CH_2]_3-CH_3$	Valeryl	285
C_5H_9S	$-CS-[CH_2]_3-CH_3$	Thiovaleryl	
C_5H_{10}	$-[CH_2]_5-$	Pentandiyl	
$C_5H_{10}NO$	$-CO-C(NH_2)(CH_3)-CH_2-CH_3$	Isovalyl	
	$-CO-CH(NH_2)-CH_2-CH_2-CH_3$	Norvalyl	
	$-CO-CH(NH_2)-CH(CH_3)_2$	Valyl	362
$C_5H_{10}NOS$	$-CO-CH(NH_2)-[CH_2]_2-S-CH_3$	Methionyl	381
	$-CO-CH(NH_2)-C(CH_3)_2-SH$	Penicillaminyl	
C_5H_{11}	$-CH_2-CH_2-CH(CH_3)_2$	Isopentyl	202
	$-CH_2-C(CH_3)_3$	Neopentyl	204
	$-CH_2-CH_2-CH_2-CH_2-CH_3$	Pentyl	201
	$-C(CH_3)-CH_2-CH_3$	*tert*-Pentyl	203

Formula	Structure	Name	Page
$C_5H_{11}N_2O$	$-CO-CH(NH_2)-[CH_2]_3-NH_2$	Ornithyl	360
$C_5H_{12}N$	$-CH_2-N(C_2H_5)_2$	Diäthylaminomethyl	
C_6H_5	$-C_6H_5$	Phenyl	229
C_6H_5N	$=N-C_6H_5$	Phenylimino	120
$C_6H_5N_2$	$-N=N-C_6H_5$	Phenylazo	15
$C_6H_5N_2O$	$-(N_2O)-C_6H_5$	Phenylazoxy	
	$-N(O)=N-C_6H_5$	Phenyl-*NNO*-azoxy	
C_6H_5O	$-O-C_6H_5$	Phenoxy	77
C_6H_5OS	$-SO-C_6H_5$	Benzolsulfinyl	132
$C_6H_5O_2S$	$-SO_2-C_6H_5$	Benxolsulfonyl	136
C_6H_5S	$-S-C_6H_5$	Phenylmercapto	99
C_6H_5S-	C_6H_5S-	Thiophenolato	
$C_6H_5S_2$	$-S-S-C_6H_5$	Phenyldisulfanyl	
C_6H_6N	$-NH-C_6H_5$	Anilino	
$C_6H_6NO_2S$	$-SO_2-NH-C_6H_5$	Phenylsulfamoyl	
	$-SO_2-C_4H_4-NH_2$	Sulfanilyl	138
$C_6H_6N_2$	$=N-NH-C_6H_5$	Phenylhydrazono	126
$C_6H_7N_2$	$-NH-NH-C_6H_5$	N′-Phenyl-hydrazino	145
C_6H_7O	$-CO-CH=CH-CH=CH-CH_3$	Hexa-2,4-dienoyl	
	$-CO-CH=CH-CH=CH-CH_3$	Sorboyl	
$C_6H_8O_2$	$-CO-CH_2-CH_2-CH_2-CH_2-CO-$	Adipoyl	314
$C_6H_{11}O$	$-CO-[CH_2]_4-CH_3$	Hexanoyl	288
$C_6H_{11}O_3$	$-CO-CH(OH)-C(CH_3)_2-CH_2-OH$	Pantoyl	
$C_6H_{11}S$	$-CS-[CH_2]_4-CH_3$	Hexanthioyl	
C_6H_{12}	$-[CH_2]_6-$	Hexandiyl	
	$=CH-[CH_2]_4-CH_3$	Hexyliden	

Formula	Structure	Prefix	Priority
$C_6H_{12}NO$	$-CO-CH(NH_2)-CH(CH_3)-C_2H_5$ (*threo*)	Alloisoleucyl	
	(*erythro*)	Isoleucyl	366
	$-CO-CH(NH_2)-CH_2-CH(CH_3)_2$	Leucyl	365
	$-CO-CH(NH_2)-[CH_2]_3-CH_3$	Norleucyl	363
C_6H_{13}	$-CH_2-[CH_2]_4-CH_3$	Hexyl	205
	$-CH_2-CH_2-CH_2-CH(CH_3)_2$	Isohexyl	206
$C_6H_{13}N_2O$	$-CO-CH(NH_2)-[CH_2]_4-NH_2$	Lysyl	364
C_7H_5	$\equiv C-C_6H_5$	Benzylidin	
C_7H_5O	$-CO-C_6H_5$	Benzoyl	299
C_7H_8OS	$-S-CO-C_6H_5$	Benzoylmercapto	
$C_7H_5O_2$	$-O-CO-C_6H_5$	Benzoyloxy	87
C_7H_5S	$-CS-C_6H_5$	Thiobenzoyl	301
C_7H_6	$=CH-C_6H_5$	Benzyliden	265
	$> CH-C_6H_5$	Benzyliden	
C_7H_6N	$-C(C_6H_5)=NH$	Benzimidoyl	
	$-C(C_6H_5)=NH$	a-Imino-benzyl	
C_7H_6NO	$-C(C_6H_5)=N-OH$	Benzohydroximoyl	
	$-NH-CO-C_6H_5$	Benzoylamino	
	$-CO-NH-C_6H_5$	Phenylcarbamoyl	334
$C_7H_6O_2$	$-O-CH(C_6H_5)-O-$	Benzylidendioxy	
C_7H_7	$-CH_2-C_6H_5$	Benzyl	232
	$-C_6H_4-CH_3$	Tolyl	231
$C_7H_7N_2$	$-C(C_6H_5)=N-NH_2$	Benzohydrazonoyl	
C_7H_7O	$-O-CH_2-C_6H_5$	Benzyloxy	79
$C_7H_7O_2S$	$-SO_2-CH_2-C_6H_5$	Phenylmethansulfonyl	
	$-SO_2-C_6H_4-CH_3$	Toluolsulfonyl	137

C_7H_7S	$-S-CH_2-C_6H_5$	Benzylmercapto	100
C_7H_8N	$-NH-CH_2-C_6H_5$	Benyzlamino	29
	$-NH-C_6H_4-CH_3$	Toluidino	28
C_7H_8NO	$-NH-C_6H_4-OCH_3$	Anisidino	30
$C_7H_{10}O_2$	$-CO-[CH_2]_5-CO-$	Heptandioyl	
$C_7H_{13}O$	$-CO-[CH_2]_5-CH_3$	Heptanoyl	
C_7H_{15}	$-CH_2-[CH_2]_5-CH_3$	Heptyl	
C_8H_7	$-CH=CH-C_6H_5$	Styryl	237
C_8H_7O	$-CH_2-CO-C_6H_5$	Phenacyl	271
C_8H_7O	$-CO-CH_2-C_6H_5$	Phenylacetyl	
	$-CO-C_6H_4-CH_3$	Toluoyl	302
$C_8H_7O_2$	$-CO-O-CH_2-C_6H_5$	Benzyloxycarbonyl	
	$-CO-CH(OH)-C_6H_5$	Mandeloyl	348
C_8H_8	$-CH_2-C_6H_4-CH_2-$	Xylylen	
C_8H_9	$-CH_2-CH_2-C_6H_5$	Phenäthyl	236
C_8H_9O	$-O-CH_2-CH_2-C_6H_5$	Phenäthyloxy	
$C_8H_{12}O_2$	$-CO-[CH_2]_6-CO-$	Octandioyl	
$C_8H_{15}O$	$-CO-[CH_2]_6-CH_3$	Octanoyl	
C_8H_{16}	$-[CH_2]_8-$	Octandiyl	
C_8H_{17}	$-CH_2-[CH_2]_6-CH_3$	Octyl	
C_9H_8	$=CH-CH=CH-C_6H_5$	Cinnamyliden	267
$C_9H_8NO_2$	$-CO-CH_2-NH-CO-C_6H_5$	Hippuroyl	357
C_9H_9	$-CH_2-CH=CH-C_6H_5$	Cinnamyl	238
$C_9H_9O_2$	$-CO-CH(C_6H_5)-CH_2-OH$	Tropoyl	
$C_9H_{10}NO$	$-CO-CH(NH_2)-CH_2-C_6H_5$	Phenylalanyl	368
$C_9H_{14}O_2$	$-CO-[CH_2]_7-CO-$	Nonandioyl	
$C_9H_{17}O$	$-CO-[CH_2]_7-CO-$	Nonanoyl	

Formula	Structure	Prefix	Priority
C_9H_{19}	$-CH_2-[CH_2]_7-CH_3$	Nonyl	
$C_{10}H_{16}O_2$	$-CO-[CH_2]_8-CO-$	Decanedioyl	
$C_{10}H_{19}O$	$-CO-[CH_2]_8-CH_3$	Decanoyl	
$C_{10}H_{21}$	$-[CH_2]_9-CH_3$	Decyl	
Cl	$-Cl$	Chloro	2
ClHN	$-NH-Cl$	Chloroanino	18
ClHOP	$-PH(O)Cl$	Chlorophosphinoyl	
ClHg	$-Hg-Cl$	Chloromercurio	
ClO	$-ClO$	Chlorosyl	
ClOS	$-SO-Cl$	Chlorosulfinyl	
ClO_2	$-ClO_2$	Chloryl	
ClO_2S	$-SO_2OCl$	Chlorosulfonyl	
ClO_3	$-ClO_3$	Perchloryl	
ClS	$-S-Cl$	Chlorosulfanyl	
ClSe	$-Se-Cl$	Chloroselanyl	
Cl_2I	$-ICl_2$	(Dichloro-1^3-iodanyl)	
$C_{12}OP$	$-POC_{12}$	Dichlorophosphoryl	
Cl_2P	$-PCl_2$	Dichlorophosphanyl	
Cr	replacement of CH by Cr	1^N-Chroma	
D	$-D(-^2H)$	Deuterio	
F	$-F$	Fluoro	1
Ge	replacement of C by Ge	Germa	
GeH_3	$-GeH_3$	Germyl	178
Ge_2H_5	$-GeH_2-GeH_3$	Digermanyl	
H	H	Hydrido	

Formula	Symbol	Name	Page
HI]$^+$	$-$IH]$^+$	Jodonio	
HN	$-$NH$-$	Azanediyl	
	$-$NH$-$	Epimino	60
	NH	Imido	
	$=$NH	Imino	115
HNO	$=$N$-$OH	Hydroxyimino	
HNO$_2$	$=$NO$-$OH	*aci*-Nitro	13
HNP	$-$P($=$NH)$-$	Phosphorimidoyl	
HN$_2$	$-$N$=$NH	Diazenyl	
HN$_2$O$_2$	$-$N(O)$=$N$-$OH	Isonitramino	
	$-$NH$-$NO$_2$	Nitroamino	
HO	OH	Hydroxo	
HO	$-$OH	Hydroxy	61
HOP	$>$PH$=$O	Phosphonoyl	
HOS	$-$S$-$OH	Hydroxysulfanyl	
HOSe	$-$Se$-$OH	Hydroxyselanyl	
HO$_2$	$-$O$-$OH	Hydroperoxy	
HO$_2$P	$>$P(O)$-$OH	Hydroxyphosphoryl	
HO$_2$S	$-$SO$-$OH	Sulfino	131
HO$_2$S$_2$	$-$SO$_2$SH	Thiosulfo	
HO$_2$Se	$-$SeO$-$OH	Selenino	139
HO$_3$S	$-$SO$_2$$-$OH	Sulfo	133
HO$_3$Se	$-$SeO$_2$$-$OH	Selenono	141
HS	$-$SH	Mercapto	94
HS$_2$	$-$S$-$SH	Disulfanyl	
HSe	$-$SeH	Selanyl	
HSe$_2$	$-$Se$-$SeH	Diselanyl	

Formula	Structure	Prefix	Priority
HTe	$-TeH$	Tellanyl	
H_2I	$-IH_2$	l_3-Jodanyl	
H_2N	NH_2	Amido	
	$-NH_2$	Amino	17
$H_2N]^+$	$=NH_2]^+$	Imonio	121
H_2NO	$-O-NH_2$	Aminooxy	
	$-NH-OH$	Hydroxyamino	143
H_2NOP	$>PO-NH_2$	Aminophosphoryl	
H_2NOS	$-SO-NH_2$	Sulfinamoyl	
H_2NO_2S	$-SO_2-NH_2$	Sulfamoyl	
H_2NO_3S	$-NH-SO_2-OH$	Sulfoamino	19
H_2NS	$-S-NH_2$	Aminosulfanyl	
	$-NH-SH$	Mercaptoamino	
H_2N_2	$>N-NH_2$	Diazane-1,1-diyl	
	$-NH-NH-$	Diazane-1,2-diyl	
	$-NH-NH-$	Epidiazano	
	$=N-NH_2$	Hydrazono	125
H_2O	H_2O	Aqua	
$H_2O]^+$	$-OH_2]^+$	Oxonio	83
H_2OP	$-PH_2=O$	Phosphinoyl	
H_2O_2P	$-PH(O)-OH$	Hydroxyphosphinoyl	159
H_2O_3P	$-PO(OH)_2$	Phosphono	160
H_2P	$-PH_2$	Phosphanyl	153
H_2PS	$-PH_2=S$	Phosphinothioyl	158
$H_2S]^+$	$-SH_2]^+$	Sulfonio	101

Formula	Group	Name	Page
H_2Sb	$-SbH_2$	Stibanyl	169
$H_2Se]^+$	$-SeH_2]^+$	Selenonio	
H_3N	NH_3	Ammine	
$H_3N]^+$	$-NH_3]^+$	Ammonio	
H_3NP	$-PH_2{=}NH$	Phosphinimidoyl	
H_3N_2	$-NH-NH_2$	Hydrazino	125
$H_3O_6P_2$	$-PO(OH)-O-PO(OH)_2$	Trihydroxydiphosphoryl	
H_3P	$=PH_3$	1^5-Phosphanylidene	
$H_3P]^+$	$-PH_3]^+$	Phosphonio	
H_3P_2	$-PH-PH_2$	Diphosphanyl	
H_3Pb	$-PbH_3$	Plumbanyl	
H_3S	$-SH_3$	1^4-Sulfanyl	
H_3Se	$-SeH_3$	1^4-Selanyl	
H_3Si	$-SiH_3$	Silyl	174
H_3Sn	$-SnH_3$	Stannanyl	179
H_3Te	$-the_3$	1^4-Tellanyl	
$H_4N_{20}P$	$-PO(NH_2)_2$	Diaminophosphoryl	
H_4N_3	$-NH-NH-NH_2$	Triazanyl	
H_4P	$-PH_4$	1^5-Phosphanyl	
$H_4S]^+$	$-SH_4]^+$	1^4-Sulfaniumyl	
H_5OSi_2	$-SiH_2-O-SiH_3$	Disiloxanyl	
H_5Si_2	$-SiH_2-SiH_3$	Disilanyl	176
Hg^+	$-Hg^+$	Mercurio[(1+)]	
I	replacement of CH by I	1^3-Joda	
	$-I$	Jodo	4
I^+	replacement of CH_2 by I^+	Jodonia	9

Formula	Structure	Prefix	Priority
IO	$-IO$	Jodoso	5
IO_2	$-IO_2$	Jodyl	10
Mg^+	$-Mg^+$	Magnesio$^{(1+)}$	
N	replacement of CH by N	Aza	
	$-N<$	Azanetriyl	
	$-N=$	Azanylylidene	
	$-N=$	Epiazanylyliden	
N	Nitrido		
	$\equiv N$	Nitrilo	
N^+	replacement of C by N^+	Azonia	
NO	$-NO$	Nitroso	11
NO	Nitrosyl		
NO_2	NO_2	Nitrito	
	$-NO_2$	Nitro	12
	$-O-NO$	Nitrosyloxy	
NO_3	NO_3	Nitrato	
	$-O-NO_2$	Nitryloxy	64
NS	$-NS$	Thionitroso	
N_2	$-N=N-$	Azo	
	$-N=N$	Diazo	128
N_2	$-N=N-$	Diazenediyl	
	$=N-N=$	Epidiazanediylidene	
	$-N=N-$	Epidiazeno	

$N_2]^+$	$-N-N]^+$	Diazonio	148
N_2O	$-N(O)=N-$	Azoxy	
	$=N-NO$	Nitrosoimino	116
N_2O_2	$=N-NO_2$	Nitroimino	
N_3	$-N_3$	Azido	14
O	$-O-$	Epioxido	
	$-O-$	Epoxy	92
	replacement of CH_2 by O^+	Oxa	
	$=O$	Oxo	114
	$-O-$ or $->O$	Oxy	
O!	$-O!$	Ylooxy	
O+	replacement of CH by O^+	Oxonia	83
O^-	O^-	Oxido	92
OP	$-PO$	Phosphoroso	155
	$>P(O)-$	Phosphoryl	
OS	$-O-S-$	Epioxidosulfano	
	$-S(O)-$	Sulfinyl	
OSe	$-Se(O)-$	Seleninyl	140
OTe	$-Te(O)-$	Tellurinyl	
O_2	$-O-O-$	Epidioxido	93
	$-O-O-$	Epidioxy	
	O_2	Peroxo	
O_2P	$-PO_2-$	Phospho	
O_2S	$-SO_2-$	Sulfonyl	
O_2Se	$-SeO_2-$	Selenonyl	142
O_3	$-O-O-O-$	Epitrioxido	

Formula	Structure	Prefix	Priority
O_3P_2	>P(O)–O–P(O)<	Diphosphoryl	
O_3S	SO_3	Sulfito	
$O_3S]-$	$-SO_3]-$	Sulfonato	
O_4S	SO_4	Sulfato	
O_5P_3	>P(O)–O–P(O)–O–P(O)<	Triphosphoryl	
O_5S_2	$-SO_2-O-SO_2-$	Disulfuryl	
Os	replacement of CH_2 by Os	1^n-Osma	
P	replacement of CH by P	Phospha	
	$-P<$	Phosphanetriyl	
	$-P=$	Phosphanylylidene	
P^+	replacement of C by P^+	Phosphonia	154
PS	>P(S)–	Thiophosphoryl	
Pb^{3+}	$-Pb^{3+}$	Plumbio$^{(3+)}$	181
S	$-S-$	Epithio	109
	$-S-$	Sulfanediyl	
	S	Sulfido	
	replacement of CH_2 by S	Thia	
	replacement of O by S	Thio	
	$=S$	Thioxo	129
S^+	replacement of CH by S^+	Thionia	
S_2	$-S-S-$	Disulfanediyl	
	S_2	Disulfido	
	$-S-S-$	Epidithio	110
S_3	$-S_3-$	Trisulfanediyl	
Sb	replacement of CH by Sb	Stiba	

Sb^{4+}	$-Sb^{4+}$	Antimonio$^{(4+)}$
Se	$-Se-$	Episelano
	$-Se-$	Selanediyl
	replacement of CH_2 by Se	Selena
Se	replacement of O by Se	Seleno
	$=Se$	Selenoxo
Se^+	replacement of CH by Se^+	Selenonia
Se_2	$-Se-Se-$	Diselanediyl
	Se_2	Diselenido
	$-Se-Se-$	Epidiselano
Si	replacement of C by S	Sila
Sn^{3+}	$-Sn^{3+}$	Stannio$^{(3+)}$
T	$-T(_3H)$	Tritio
Te	replacement of CH_2 by Te	Tellura
	replacement of O by Te	Telluro

B.4.

Chemical Abstracts:

QD1.A51

- compound name and structure
- preparation and derivatives
- physical and chemical properties
- analytical and spectroscopic information
- patent references

B.4.a. OVERVIEW

Chemical Abstracts publishes abstracts of all scientific and technical papers, including patents, conference and symposium proceedings, books, government reports, and dissertations of chemical interest appearing in the world's chemical literature. The focus of the Abstracts is to index every compound that has been isolated or synthesized.

EASY ACCESS

— For a chemical, look in the most recent Formula Index for the preferred Chemical Abstracts index name, then proceed to the Chemical Substance Index. (Beware of nomenclature changes before Volume 76 (1972). Check the Formula Index of Collective Volume 8 or the Index Guide.)

INDEXING

Each weekly issue of **Chemical Abstracts** contains Author, Key word, and Patent Indexes. Each year two volumes are issued with Author, Chemical Substance, Formula, General Subject, Index of Ring Systems, and Patent indexes. Every five years a cumulative index is published.

The **Author Index** [Collective Indexes 1–Present] links the names of authors, patentees, and patent assignees together with paper or patent titles to the corresponding Chemical Abstract Number or accession number. All names of coauthors are included at the name of the first-listed author. The rules and conventions for the spelling and transliteration of authors' names are explained in the introduction to the index.

The **Chemical Substance Index** [Collective Indexes 9–Present] lists the chemical substances with their registry numbers in square brackets. The chemical names are inverted with the parent compound followed by its derivatives. A short descriptor of the substance's context or relationship in the original document and its corresponding abstract number is given. *Beware*: A substance cited before Volume 76 (1972) may be found under its trivial name rather than under the currently employed systematic nomenclature.

The **Formula Index** [Collective Indexes 4–Present] lists the molecular formula of chemical substances with their registry numbers in square brackets. For common substances, cross-references direct one to the Chemical Substance Index; otherwise a complete set of references is provided. Formulas are arranged by citing first the number of carbons, then hydrogens, and finally the remaining elements in alphabetical order by element symbol. If carbon is not present, the elements are arranged in alphabetical order. The formulas of the parent substances are used for hydrochloride, acetate, etc., and salts of amines or metal salts of carboxylic acids or alcohols.

The **General Subject Index** [Collective Indexes 9–Present] contains all entries that do not refer to specific chemical substances. This includes classes of compounds, reactions, manufacturing apparatus and processes, general biochemical and biological topics, and incompletely defined substances.

The **Index Guide** [Collective Indexes 8–Present] cross-references common chemical names and terms with the preferred Chemical Abstracts index name. This index leads to specific entries in the Chemical Substance Index and General Subject Index for trivial natural product names, pharmaceutical names, trade names, enzymes, mixtures, and alloy names. Cross-references for substance names used as index headings before Volume 76 (1972) are also reported.

The **Numerical Patent Index** [Collective Indexes 4–9] links the country and specification number of each patent with its abstract number. The patents are sequenced in alphabetical order by country in ascending numerical order. For inventions patented in more than one country, an abstract is provided only for the patent first received by **Chemical Abstracts**.

The **Patent Index** [Collective Indexes 10–Present] relates patent numbers to abstract numbers. The patents numbers are arranged alphabetically by country code (see: B.4.d. Patent Country Codes) and are listed in ascending numerical order. Each entry also contains a code describing the type of document and all related or equivalent patents. Any patent not related to an abstract is cross-referenced to the equivalent patent number where patent family information can be found.

The **Patent Concordance** [Collective Indexes 7–9] correlates patents issued by different countries with all equivalent patents in the patent family. The patents are sequenced in alphabetical order by country in ascending numerical order. Each entry is referenced to all corresponding patents in other countries and the pertinent abstract number.

The **Index of Ring Systems** [Collective Indexes 6–Present] lists the names of cyclic and heterocyclic compounds in the order determined by their ring analyses. The arrangement of the Ring System Index is based on a type of fragmentation code: (1) the number of rings in the compound, i.e., 2-RING SYSTEMS or 3-RING SYSTEMS; (2) the size of the individual rings in the compound, such as 6,6 or 5,6,6; and (3) the elemental analysis of the individual rings in the compound with the carbon atoms listed first followed by the other elementals in alphabetical order. The elemental analyses are grouped by their component rings and listed in the same order as the ring sizes in (2), for example, $C_4N_2-C_5N$ or $C_5-C_4N_2-C_5N$.

The **Subject Index** [Collective Indexes 1–8] lists chemical substances and general subjects that do not refer to specific chemical compounds with a descriptor and abstract reference. The chemical names have their registry numbers in square brackets, and the parent compound is listed first followed by its derivatives. The general subjects include classes of compounds, general chemical reactions, laboratory and manufacturing apparatus, analytical and physical instrumentation, and general biochemical and biological subjects.

COMPUTER SEARCH

Chemical Abstracts databases are available through STN International by typing FILE CA for abstracts published after 1967 and FILE CAOLD for abstracts before 1967. It is available through Dialog Information Retrieval Service database [399]. STN has a variety of other scientific databases available (see: B.4.c. CAS Online File Names). Information can be assessed by searching the desired Search Field; aids are available by typing HELP or in the STN Database Summary Sheet. The LCA (Learning CA) File is a training database for learning how to use the CA and CApreviews files. Example search fields and display codes are given below. A more accurate search can be conducted using Roles found in section B.4.b.

Search Field Name	Search Code	Search Example	Display Code
Accession Number	/AN	S 102:123456	
Author	/AU	S CONNERS/AU	References
Beilstein Registry Number	/BRN	S 130733/BRN	BRN
CAS Registry Number	/RN	S 89123-39-7/RN	RN

Chemical Name	/CN	S BENZENE/CN	CN
Molecular Formula	/MF	S C3H5/MF	MF
Patent Number	/PN	S 5484983/PN	References

The default display format, DISPLAY L(*ist number*), in CA provides information on the accession number (AN), title of document (TI), author names (AU), corporate source (CS), literature source (SO), document type (DT), and the document's language (LA). The DISPLAY ALL command also gives the classification code [CA section and cross-references] (CC), graphic image (GI), abstract text (AB), supplemental terms [key words] (ST), and index terms (IT). The CAOLD file provides only the accession number (AN) and the index terms (IT). Any of these individual data fields can be accessed with the DISPLAY command.

(Copyright 1998 by the American Chemical Society and reprinted with permission.)

INTERNET

Chemical Abstracts Service web page is: http://info.cas.org. The page offers information about CAS, products and services, support and training, and new products. There is also access to an introductory online search tutorial using CAS databases on STN.

STN Database Summary Sheets can be accessed at:

http://www.cas.org/ONLINE/DBSS/dbsslist.html

Dialog Information Retrieval Service Blue Sheets can be accessed at:

http://www.dialog.com/bluesheets/

The Clearinghouse for Chemical Information Instructional Materials at:

http://www.indiana.edu/~ cheminfo/cciim31.html

contains tips, tutorials, and guides to various aspects of Chemical Abstracts.

CD-ROM

Chemical Abstracts on CD-ROM contains all the information available in the printed edition. It can be searched by bibliographic data, abstracts, abstract number, substance names, or molecular formula. The 13th and 12th Collective Indexes and CASSI are also available on CD-ROM.

LITERATURE DESCRIPTION

Hedda Schultz, *From CA to CAS Online*; VCH Publishers: New York, 1988 (O25.0654).

ORGANIZATION

Each volume of **Chemical Abstracts** covers one half year in 26 weekly issues. The abstracts are divided into five subject groupings and 80 sections; the Biochemistry (1–20) and Organic Chemistry (21–34) groupings appear in the odd weeks, while Macromolecular (35–46), Applied Chemistry and Chemical Engineering (47–64), and Physical and Analytical Chemistry (65–80) appear in the even weeks. A complete index of the section titles can be found in B.4.e. Three types of source documents are identified by capitalized code letters preceding the reference; these code letters are B — for books, P — for patents, and R — for reviews.

The abstracts are numbered serially throughout each volume, with each number followed by a letter to serve as a check character for the prevention of errors. From Volume 66 (1964) to volume 28 (1934), only the columns were numbered and each column was divided into nine sections, labeled a to i (or ^{1} to ^{9} before Volume 41). Each abstract was referred to by its column number and a letter (or superscript number) indicating the fraction of the column in which it started.

GUIDES

The **Registry Handbook — Common Names** (Microfilm) links the common names for substances to their CAS Registry Numbers, CA index names, and molecular formulas.

The **Registry Handbook — Number Section** relates Registry Numbers to their corresponding chemical substances and molecular formulas. The presence of an asterisk (*) following the Registry Number indicates that the substance is not a unique chemical entity and can only be found either in the General Subject Index or as an incompletely defined derivative of a specific chemical substance in the Chemical Substance Index.

The **Chemical Abstracts Service Source Index (CASSI)** contains a complete listing of every serial abstracted in **Chemical Abstracts**. The abbreviation of each title is indicated in boldface (**J**ournal of the **Am**erican **Chem**ical **Soc**iety) and arranged alphabetically letter-by-letter according to the abbreviated form of the serial title (JAMCHEMSOC).

The Source Index also lists those libraries that have the serial in their collection and relevant publication information. The entry "Doc. Supplier: CAS" indicates that the title may be available from the CAS Document Delivery Service (call 800-678-4337 or 614-447-3670 from 7:00 A.M. to 7:30 P.M. Eastern Standard Time, Monday through Friday; fax 614-447-3648; or email to dds@cas.org).

EXAMPLE ENTRY

Journal

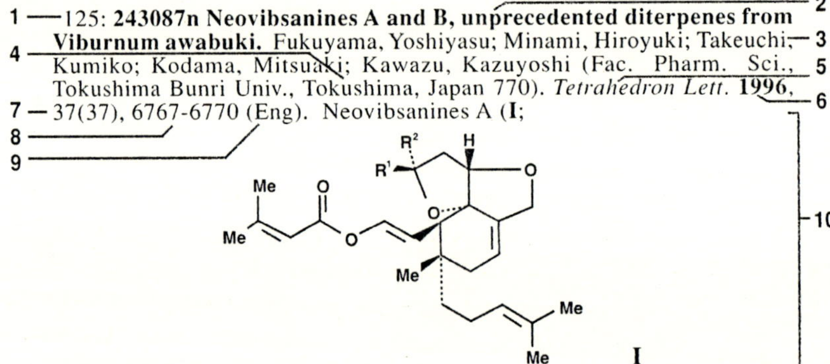

R^1 = OMe, R^2 = Me) and B **I** (R^1 = Me, R^2 = OMe), isolated from the leaves of Viburnum awabuki, are unprecedented tricyclic diterpenoids, which are presumably biosynthesized via consecutive ring cleavage and cyclization from vibsanine B. Their relative structures have been established by extensive anal. of spectroscopic data and photochem. reaction of vibsanine B.

1.	CA volume and abstract number	7.	volume and issue of publication
2.	document title	8.	inclusive pagination
3.	author name(s)	9.	language of document
4.	address where the work was conducted	10.	abstract, often including structure diagrams
5.	abbreviated journal title		
6.	year of publication		

B.4.b.   CHEMICAL ABSTRACTS ROLES

Chemical Abstracts Roles are used to narrow a search in order to obtain more precise information on substances. Every indexed substance has been assigned roles based on scientific analysis to the original indexed document. Roles can be searched in CA, HCA, CAplus, and HCAplus files.

Super roles have 4-letter codes. Specific roles have 3-letter codes.

Copyright 1998 by the American Chemical Society and reprinted with permission.

Analytical Study	**ANST**
Analyte	ANT
Analytical Matrix	AMX
Analytical Reagent Use*	ARG
Analytical Role, Unclassified	ARU
Biological Study	**BIOL**
Adverse Effect, Including Toxicity	ADV
Agricultural Use*	AGR
Bioindustrial Manufacture*	BMF
Biological Activity or Ejector, Except Adverse	BAC
Biological Occurrence	BOC
Biological Process*	BPR
Biological Use, Unclassified	BUU
Biological Study, Unclassified	BSU
Biosynthetic Preparation*	BPN
Food or Feed Use*	FFD
Metabolic Formation*	MFM
Therapeutic Use*	THU
Formation, Nonpreparative	**FORM**
Formation, Unclassified	FMU
Geological or Astronomical Formation	GFM
Metabolic Formation*	MFM
Occurrence	**OCCU**
Biological Occurrence*	BOC
Geological or Astronomical Occurrence	GOC
Occurrence, Unclassified	OCU
Pollutant	POL

Preparation   **PREP**
Bioindustrial Manufacture*   BMF
Biosynthetic Preparation*   BPN
Byproduct   BYP
Industrial Manufacture   IMF
Purification   PUR
Preparation, Unclassified   PNU
Synthetic Preparation   SPN

PROC Process
Biological Process*   BPR
Geological or Astronomical Process   GPR
Physical, Engineering, or Chemical Process   PEP
Removal or Disposal   REM

Uses
Agricultural Use*   AGR
Analytical Reagent Use*   ARG
Biological Use, Unclassified*   BUU
Catalyst Use   CAT
Device Component Use   DEV
Food or Feed Use*   FFD
Modifier or Additive Use   MOA
Nonbiological Use, Unclassified   NUU
Polymer in Formulation   POF
Technical or Engineered Material Use   TEM
Therapeutic Use*   THU

The following are specific roles that are not upposted to any super
roles:

Miscellaneous   MSC
Properties   PRP
Reactant   RCT

B.4.c.   CHEMICAL ABSTRACTS ONLINE FILE NAMES

Copyright 1998 by the American Chemical Society and reprinted with permission.

To look at detailed information about a file, first access that file using the FILE command. Enter HELP CONTENT at an arrow prompt ($\Rightarrow$) for a general description of the file. Enter HELP DIRECTORY for a list of help messages available for that file. The database summary sheet is also available for the file in STNGUIDE. Enter FILE STNGUIDE at an arrow prompt ($\Rightarrow$), then search the file name in the /DBN search field. You can then display the search fields, display fields, file content, sources, etc.

All files are available for multifile searching except HOME, NUMERIGUIDE, STNGUIDE, STNMAIL, and the Learning Files.

The following files are available:

A

ABC der Deutschen Wirtschaft, German Manufacturers	ABCD
ABC Europe Production Europex	EUROPEX
ABI-INFORM 1971–present	ABI-INFORM
Aeroplus Access Database 1962–present	AEROSPACE
AGRICulture OnLine Access 1970–present	AGRICOLA
AIChE Design Inst. Physical Property Data File	DIPPR
AIDS Literature Online	AIDSLINE
AIDSLINE — AIDS Literature Online	AIDSLINE
Aluminum Association Standards and Data	AAASD
Aluminum Fracture Toughness Data Base	ALFRAC
American Chemical Society Journals	CJACS
Analytical Abstracts	ANABSTR
AOAC, Chemical Journals	CJAOAC
API Literature File 1964–present (Supporters)	APILIT
API Literature File 1964–present (Non-Supporters)	APILIT2
API Patent File 1964–present (Supporters)	APIPAT
API Patent File 1964–present (Non-Supporters)	APIPAT2
Aquatic Sciences & Fisheries Abstracts 1978–present	AQUASCI

B

BEILSTEIN File of Organic Compounds	BEILSTEIN
BEILSTEIN Learning File	LBEILSTEIN
Bibliodata learning File	LBIBLIO
BioBusiness/RN 1985–present	BIOBUSINESS
BIOSIS Previews(R)/RN File 1969–present	BIOSIS
Biotechnology Abstracts 1982–present (Subscribers)	BIOTECHDS
Biotechnology Abstracts 1982–present	BIOTECHABS
BMFT Foerderkatalog	FORKAT
Business Opportunities Service, International	BUSI

C

CAB ABSTRACTS 1973–present	CABA
Cancer Literature Online 1963–present	CANCERLIT
Catalogue of Journals in German Libraries	ZDB
Catalogue of Technical Information Library in Hannover	TIBKAT
Catalogs of Chemicals 1993–1995	CHEMCATS
CBNB — Chemical Business NewsBase 1985–present	CBNB
CCOHS Material Safety Data Sheets	MSDS-CCOHS
Ceramic Abstracts from 1976	CERAB
Chem. Eng. and Biotech. Abstracts 1975–present	CEABA
Chemical Abstracts, Descriptive information about STN databases	STNGUIDE
Chemical Abstracts File (pre-1967)	CAOLD
Chemical Abstracts File — Pre-1967 — with hour-based pricing	HCAOLD
Chemical Abstracts File for 1967–present	CA
Chemical Abstracts File for 1967–present with hour-based pricing	HCA
Chemical Abstracts File for 1967–present zero connect hour	ZCA
Chemical Abstracts Learning File	LCA
Chemical Abstracts Patent Markush File 1988–present	MARPAT
Chemical Abstracts Patent Markush Learning File	LMARPAT
Chemical Abstracts Patent Markush Preview File	MARPATPREV

Chemical Abstracts Plus File for 1967–present   CAPLUS
Chemical Abstracts Plus File for 1967–present
  (hour-based pricing)   HCAPLUS
Chemical Abstracts Plus File for 1967–present
  zero connect hour   ZCAPLUS
Chemical Abstracts Reaction Search Service   CASREACT
Chemical Abstracts Reaction Search Service
  Learning File   LCASREACT
Chemical Abstracts Registry File of
  substances   REGISTRY
Chemical Abstracts Registry File of
  substances, zero connect hour   ZREGISTRY
Chemical Abstracts Registry Learning File   LREGISTRY
Chemical Business NewsBase 1985–present   CBNB
Chemical Catalogs Online 1993–1995   CHEMCATS
Chemical & Engineering News 1990–present   CEN
Chemical & Environmental Compatibility of
  Plastics (PDL)   PDLCOM
Chemical Industry Notes File for
  1974–present   CIN
Chemical Journals of Elsevier Science
  Publishers B.V.   CJELSEVIER
Chemical Journals of John Wiley & Sons, Inc.   CJWILEY
Chemical Journals of the American Chemical
  Society   CJACS
Chemical Journals of the AOAC   CJAOAC
Chemical Journals of the Royal Society of
  Chemistry   CJRSC
Chemical Journals of VCH Verlagsgesellschaft   CJVCH
Chemical Journals Online, Learning File   LCJO
Chemical safety information   CHEMSAFE
Chemical Safety News Base from
  1981–present   CSNB
Chemicals, ChemSources — USA and
  International   CSCHEM
Chemicals Listing, Regulated — with
  hour-based pricing   HCHEMLIST
Chemicals Listing, Regulated   CHEMLIST
CHEMINFORMRX Reaction Search Service   CHEMINFORMRX
CHEMREACT Reaction Search System   CHEMREACT
CHEMSAFE — chemical safety information   CHEMSAFE
ChemSources — USA and International
  (Company Directory)   CSCORP

ChemSources — USA and International
   (Chemicals)   CSCHEM
CIN File for 1974–present with hour-based
   pricing   HCIN
Coatings, Surface, World Abstracts
   1976–present   WSCA
Company Directory, ChemSources — USA
   and International   CSCORP
COMPENDEX*PLUS File 1970–present   COMPENDEX
Computer & Information Systems Abstracts
   1981–present   COMPUAB
Computerscience 1972–present   COMPUSCIENCE
Conference Papers Index 1973–present   CONFSCI
Conferences in Energy, Physics, Mathematics
   etc.   CONF
Copper and Copper Alloy Standards and Data   COPPERDATA
Cosmetic and Perfume Science & Technology   KOSMET
CRC HANDBOOK of DATA on ORGANIC
   COMPOUNDS, 2nd Edition   HODOC
Creep and Rupture Strengths of Metals   METALCREEP
Crop Protection File 1968–84, Derwent   CROPB
Crop Protection File 1985–present, Derwent   CROPU
Crop Protection Registry, Derwent   CROPR
CROPB Derwent Crop Protection File
   1968–84   CROPB
CROPU Derwent Crop Protection File
   1985–present   CROPU
CSA Life Sciences Collection 1978–present   LIFESCI

D

Default login file. Contains no data.   HOME
Derwent Crop Protection File 1968–84   CROPB
Derwent Crop Protection File 1985–present   CROPU
Derwent Crop Protection Registry   CROPR
Derwent Drug File, Backfile 1964–82
   (Subscriber)   DRUGB
Derwent Drug File 1983–present (Subscriber)   DRUGU
Derwent Drug File 1983–present
   (Non-subscriber)   DDFU
Derwent Drug File, Backfile 1964–82
   (Non-subscriber)   DDFB
Derwent Drug Learn File   LDRUG
Derwent Geneseq Database 1981–present   DGENE

Derwent Patents Citation Index 1978–present   DPCI
Derwent Patents Citation Index Learning File   LDPCI
Derwent Reaction Search Service DJSM for
   subscriber   DJSMDS
Derwent Reaction Search Service DJSM   DJSMONLINE
Derwent Veterinary Drug File 1968–82
   (subscribers)   VETB
Derwent Veterinary Drug File 1983–present   VETU
Descriptive information about STN databases   STNGUIDE
DETHERM-DECHEMA thermophysical
   property database   DETHERM
Dissertation Abstracts 1861–present   DISSABS
DMSworld Drug Updates 1977–present   DRUGUPDATES
DOE ENERGY file 1974–present   ENERGY
Drug development, Pharmaprojects, status file   PHAR
Drug File, Derwent, Backfile 1964–82
   (Subscriber)   DRUGB
Drug File, Derwent, from 1983–present
   (Subscriber)   DRUGU
Drug File, Derwent, Backfile 1964–82
   (Non-subscriber)   DDFB
Drug File, Derwent, 1983–present
   (Non-subscriber)   DDFU
Drug Monographs, IMSworld, 1960s–present
   (with prices)   DRUGMONOG
Drug Monographs, IMSworld, 1960s–present
   (no prices)   DRUGMONOG2
Drug Launches, IMSworld, 1982–present   DRUGLAUNCH
Drug Learn File, Derwent   LDRUG
Drug News, IMSworld, 1991–present   DRUGNL
Drug Patents International, IMSworld,
   1987–present   DRUGPAT
Drug Updates, DMSworld, 1977–present   DRUGUPDATES

E
East German Patents from 1982–present   PATDD
EC research projects in energy, 1975–present   SESAME
Educational and Research Institutions, F.R.
   Germany   VADEMECUM
Electronics & Communications Abstracts
   1981–present   ELCOM
Elsevier Science Publishers B.V., Chemical
   Journals   CJELSEVIER

EMBASE Alert	EMBAL
EMBASE File from 1974–present	EMBASE
EMBASE Learning File	LEMBASE
Energy Conferences	CONF
ENERGY file, DOE, from 1974–present	ENERGY
Energy file, FIZ Karlsruhe, from 1976–present	ENERGIE
Engineered Materials Abstracts File from 1986–present	EMA
Environmental Literature from 1976–present	ULIDAT
Environment Research in Progress from 1974–present	UFORDAT
European (EP) Patents full text from 1996–present	EUROPATFULL
European Patent Office Patents from 1978–present	PATOSEP
Europe Production Europex, ABC	EUROPEX
Europex, ABC Europe Production	EUROPEX
EVENTLINE- Events worldwide 1989–2000+	EVENTLINE
Events worldwide 1989–2000+	EVENTLINE

F

FIZ Karlsruhe Energy file from 1976–present	ENERGIE
Food Science Technology Abstracts from 1969–present	FSTA

G

Genetic Sequence Data Bank	GENBANK
Geneseq Database, Derwent, 1981–present	DGENE
Geological Reference File from 1785–present	GEOREF
GeoRef — Geological Reference File from 1785–present	GEOREF
German Libraries, Catalogue of Journals	ZDB
German literature in social sciences 1945–present	SOLIS
German Manufacturers: ABC der Deutschen Wirtschaft	ABCD
German National Bibliography from 1972–present	BIBLIODATA
German Patent Database from 1968–present	PATDPA
German Patents and Utility Models from 1968–present	PATOSDE

Global Mobility Database from 1906–present	1MOBILITY
Global Mobility Standards Database	2MOBILITY
Gmelin Handbook of Inorganic Chemistry	GMELIN
Government Reports Announcements 1974–present	NTIS
Grey Literature in Europe from 1981–present	SIGLE

H

Hazardous Substances Databank	HSDB
Health and Safety Science Abstracts 1981–present	HEALSAFE
HOME. The default login file. Contains no data	HOME

I

ICSD — Inorganic Crystal Structure Data File	ICSD
IFI Comprehensive Database from 1950–present	IFICDB
IFI Patent Database from 1950–present	IFIPAT
IFI Reassignment and Reexamination Database	IFIRXA
IFI Uniterm Database from 1950–present	IFIUDB
IFI Uniterm and U.S. Class Reference File	IFIREF
IMSworld Drug Launches from 1982–present	DRUGLAUNCH
IMSworld Drug Monographs 1960s–present (with prices)	DRUGMONOG
IMSworld Drug Monographs 1960s–present (no prices)	DRUGMONOG2
IMSworld Drug News from 1991–present	DRUGNL
IMSworld Drug Patents International from 1987–present	DRUGPAT
Information Science and Work from 1976 to present	INFODATA
Inorganic Crystal Structure Data File	ICSD
INPADOC Bibliographic, Family, Legal Status from 1968	INPADOC
INPADOC file for SDI	INPAMONITOR
Inspec Phys Supplement Backfile (1979–1994)	INSPHYS
INSPEC, Learning File	LINSPEC
INSPEC2 File from 1969–present	INSPEC
INSPHYS — Inspec Phys Supplement Backfile (1979–1994)	INSPHYS
Institutes and Research in Progress	INFOR
International Business Opportunities Service	BUSI

International Construction Database from
    1976–present   ICONDA
International Nuclear Information System
    1970–present   INIS
International Pharmaceutical Abstracts
    1970–present   IPA
International Plastics Selector   IPS
INVESTEXT from 1982 to present   INVESTEXT
ISMEC — Mechanical Engineering Abstracts   ISMEC

J

Japanese Government and Public Research in
    Progress   JGRIP
Japanese Patents from 1976–present   JAPIO
JAPIO — Japanese Patents from 1976–present   JAPIO
JICST-EPlus File on Sci. & Tech. in Japan
    1985–present   JICST-EPLUS
John Wiley & Sons, Inc., Chemical Journals   CJWILEY
Journals in German Libraries, Catalogue   ZDB
Journals of Elsevier Science Publishers B.V.,
    Chemical   CJELSEVIER
Journals of John Wiley & Sons, Inc., Chemical   CJWILEY
Journal of Synthetic Methods   DJSMDS
Journals of the American Chemical Society   CJACS
Journals of the AOAC   CJAOAC
Journals of the Royal Society of Chemistry   CJRSC
Journals of VCH Verlagsgesellschaft,
    Chemical   CJVCH

L

Learn File, Derwent Drug   LDRUG
Learning File, BEILSTEIN   LBEILSTEIN
Learning File, Bibliodata   LBIBLIO
Learning File, Chemical Abstracts   LCA
Learning File, Chemical Journals Online   LCJO
Learning File, Chemical Abstracts Patent
    Markush   LMARPAT
Learning File, Chemical Abstracts Reaction
    Search Service   LCASREACT
Learning File, Derwent Patents Citation Index   LDPCI
Learning File, EMBASE   LEMBASE

Learning File, MEDLINE   LMEDLINE
Learning File, Registry   LREGISTRY
Learning File, World Patent Index   LWPI
Learning INSPEC File   LINSPEC
Life Sciences Collection (CSA) from
    1978–present   LIFESCI
LINGUISTIC LITERATURE from   BLLDB
    1971–present

M

Mail, STN Electronic Service   STNMAIL
Markush File, Chemical Abstracts Patents
    1988–present   MARPAT
Markush File, Chemical Abstracts Patent,
    Preview File   MARPATPREV
Material Safety Data Sheets (CCOHS)   MSDS-CCOHS
Material Safety Data Sheets (OHS)   MSDS-OHS
Materials Business File from 1983–present   MATBUS
Materials Database   ASMDATA
Mathematics Conferences   CONF
Mechanical Engineering Abstracts (ISMEC)   ISMEC
MEDLINE File from 1966–present   MEDLINE
MEDLINE Learning File   LMEDLINE
Merck Index Online (SM)   MRCK
METADEX File from 1966–present   METADEX
Metals, Creep and Rupture Strengths   METALCREEP
Metals Datafile   MDF
Mobility Database, Global, from 1906–present   1MOBILITY
Mobility Standards Database, Global   2MOBILITY

N

Natural Products Alert Database   NAPRALERT
Newsletter Database from 1988–present   NLDB
NIOSHTIC 1973–present   NIOSHTIC
Numeric Files, Property Hierarchy and   NUMERIGUIDE
    Directory

O

Oceanic Abstracts from 1964–present   OCEAN
OCLC Union Catalog (of library holdings)   WORLDCAT
OHS Material Safety Data Sheets   MSDS-OHS

P

Packaging Science Technol. Abstracts 1981–present	PSTA
PAPERCHEM 1967–present (IPST Member Version)	PAPERCHEM
PAPERCHEM 1967–present (Public Version)	PAPERCHEM2
Patent Database, German, from 1968–present	PATDPA
Patent Database, IFI, from 1950–present	IFIPAT
Patent Index, World, Learning File	LWPI
Patent Markush File, Chemical Abstracts, 1988–present	MARPAT
Patent Markush File, Chemical Abstracts, Preview File	MARPATPREV
Patent Markush Learning File, Chemical Abstracts	LMARPAT
Patents and Utility Models, German, from 1968–present	PATOSDE
Patents (PCT) from 1983–present	PATOSWO
Patents Citation Index, Derwent, 1978–present	DPCI
Patents Citation Index, Derwent, Learning File	LDPCI
Patents, East German, from 1982–present	PATDD
Patents, European (EP), full text from 1996–present	EUROPATFULL
Patents of European Patent Office from 1978–present	PATOSEP
Patents, Japanese, from 1976–present	JAPIO
Patents International, IMSworld Drug, from 1987–present	DRUGPAT
Patents, U.S., from 1971–present	USPATFULL
Patents, World Patents Index, 1963–present (Subscribers)	WPIDS
Patents, World Patents Index, 1963–present	WPINDEX
PCT Patents from 1983–present	PATOSWO
PDL Chemical & Environmental Compatibility of Plastics	PDLCOM
Perfume and Cosmetic Science & Technology	KOSMET
Petroleum Abstracts 1965–present	TULSA
Petroleum Abstracts 1965–present (Non-subscribers)	TULSA2
Pharmaceutical & Healthcare Industry Names (Archived)	PHIN
Pharmaceutical & Healthcare Industry News (Current)	PHIC

Pharmaceutical News Index from
   1975–present   PNI
Pharmaprojects drug development status file   PHAR
Physical Property Data File, AIChE Design   DIPPR
   Inst.
Physics Conferences   CONF
PIRA Database from 1975   PIRA
PLASPEC Daily News File   PLASNEWS
Plastics Materials Selection Database   PLASPEC
Plastics Rubber Fibres File from 1973–present   KKF
Pollution Abstracts from 1970–present   POLLUAB
PROMT from 1978–present   PROMT
Property Hierarchy and Directory for Numeric
   Files   NUMERIGUIDE
PSTA — Packaging Science Technol. Abstracts
   1981–present   PSTA

R
RAPRA — Rubber, Plastics, Polymer
   Composites 1972–present   RAPRA
Reaction Search Service, Chemical Abstracts   CASREACT
Reaction Search Service, CHEMINFORMRX   CHEMINFORMRX
Reaction Search System, CHEMREACT   CHEMREACT
Reaction Search Service, Derwent, JSM for
   subscriber   DJSMDS
Reaction Search Service, Derwent, DJSM   DJSMONLINE
Regional planning and building construction   RSWB
Registry of Toxic Effects of Chemical   RTECS
   Substances
Regulated Chemicals Listing   CHEMLIST
Regulated Chemicals Listing with hour-based
   pricing   HCHEMLIST
Registry File of substances, Chemical   REGISTRY
   Abstracts
Registry Learning File   LREGISTRY
Reports in Science and Technology from F.R.
   Germany   FTN
Research in Social Sciences from
   1980–present   FORIS
Road and Transport Information Database   IRRD
Royal Society of Chemistry, Journals   CJRSC

Rubber, Plastics, Polymer Composites
    1972–present   RAPRA
Russian Scientific News   RUSSCI

S

Science and Technology Reports from F.R.
    Germany   FTN
Science & Tech. in Japan, JICST-EPlus File
    1985–present   JICST-EPLUS
Science Citation Index   SCISEARCH
SCISEARCH from 1974–present   SCISEARCH
Social Sciences, German literature in,
    1945–present   SOLIS
Social Sciences Research from 1980–present   FORIS
Solid State and Superconductivity Abstracts
    from 1981   SOLIDSTATE
Spectral Database Information System   SPECINFO
STN databases, Descriptive information about   STNGUIDE
STN Electronic Mail Service   STNMAIL
Surface Coatings Abstracts, World,
    1976–present   WSCA
Synthetic Methods, Journal of   DJSMDS
Swets Table of Contents from 1993–present   SWETSCAN

T

Technical Information Library in Hannover,
    Catalogue of   TIBKAT
Technology Assessment Database   TA
Theilheimer   DJSMDS
Thermodynamic Research Center
    Thermodynamic Tables   TRCTHERMO
Thermophysical property database,
    DETHERM-DECHEMA   DETHERM
Toxic Effects of Chemical Substances
    (Registry)   RTECS
Toxicology Literature from Special Sources   TOXLIT
Toxicology Literature Online 1965-present   TOXLINE
TOXLINE — Toxicology Literature Online
    1965–present   TOXLINE
TOXLIT — Toxicology Literature from
    Special Sources   TOXLIT

Transport and Road Information Database   IRRD
TRIBOLOGY INDEX (Friction, Wear,
   Lubrication) 1972–present   TRIBO

U
USAN—United States Adopted Names   USAN
United States Adopted Names (USAN)   USAN
U.S. Patents from 1971–present   USPATFULL

V
VCH Verlagsgesellschaft, Chemical Journals   CJVCH
Verfahrenstechnische Berichte from
   1966–present   VTB
VETB Derwent Veterinary Drug File 1968–82
   (subscribers)   VETB
Veterinary Drug File (Derwent) 1983–present   VETU
Veterinary Drug File (Derwent) 1968–82
   (subscribers)   VETB
VETU Derwent Veterinary Drug File
   1983–present   VETU

W
World Patents Index, 1963–present
   (Subscribers)   WPIDS
World Patents Index, 1963–present   WPINDEX
World Patent Index Learning File   LWPI
World Surface Coatings Abstracts
   1976–present   WSCA
WPI—World Patents Index, 1963–present
   (Subscribers)   WPIDS
WPI—World Patents Index, 1963–present   WPINDEX

Z
Zentralblatt fuer Didaktik der Mathematik
   1976–present   MATHDI
Zentralblatt fuer Mathematik from   MATH
   1972–present
Zero connect hour for Chemical Abstracts
   File, 1967–present   ZCA
Zero connect hour for Chemical Abstracts
   Plus File 1967–present   ZCAPLUS
Zero connect hour for CA Registry File of
   substances   ZREGISTRY

B.4.d. CHEMICAL ABSTRACTS PATENT COUNTRY CODES

Copyright 1998 by the American Chemical Society and reprinted with permission.

By Code

AT	Austria
AU	Australia
BE	Belgium
BR	Brazil
CA	Canada
CH	Switzerland
CN	China, People's Republic of
CS	Czechoslovakia
CZ	Czech Republic
DD	Germany
DE	Germany
DK	Denmark
EP	European Patent Organization
ES	Spain
FI	Finland
FR	France
GB	United Kingdom
HU	Hungary
IL	Israel
IN	India
JP	Japan
LT	Lithuania
LV	Latvia
NL	Netherlands
NO	Norway
PL	Poland
RO	Romania
RU	Russia
SE	Sweden
SK	Slovakia
SU	Union of Soviet Socialist Republics
US	United States of America
WO	World Intellectual Property Orgn.
ZA	South Africa

By Country

Australia	AU
Austria	AT
Belgium	BE
Brazil	BR
Canada	CA
China, People's Republic of	CN
Czech Republic	CZ
Czechoslovakia	CS
Denmark	DK
European Patent Orgn.	EP
Finland	FI
France	FR
Germany	DD
Germany	DE
Hungary	HU
India	IN
Israel	IL
Japan	JP
Latvia	LV
Lithuania	LT
Netherlands	NL
Norway	NO
Poland	PL
Romania	RO
Russia	RU
Slovakia	SK
South Africa	ZA
Spain	ES
Sweden	SE
Switzerland	CH
Union of Soviet Socialist Republics	SU
United Kingdom	GB
United States of America	US
World Intellectual Property Orgn.	WO

B.4.e.   CHEMICAL ABSTRACTS SECTIONS

The chemical literature is divided into 80 sections. Each section covers one broad subject area. Each abstract in Chemical Abstracts appears in only one section. If there is information in the abstract that touches on another section, the abstract is cross-referenced to that section.

In Part B.4.e.1, the Chemical Abstracts Sections are listed by broad subject areas in numerical order. Part B.4.e.2 is an Index to the Chemical Abstracts Sections that alphabetically lists the section names followed by the corresponding section number.

B.4.e.1.   CHEMICAL ABSTRACTS SECTIONS

Biochemistry Sections

1. Pharmacology
2. Mammalian Hormones
3. Biochemical Genetics
4. Toxicology
5. Agrochemical Bioregulators
6. General Biochemistry
7. Enzymes
8. Radiation Biochemistry
9. Biochemical Methods
10. Microbial, Algal, and Fungal Biochemistry
11. Plant Biochemistry
12. Nonmammalian Biochemistry
13. Mammalian Biochemistry
14. Mammalian Pathological Biochemistry
15. Immunochemistry
16. Fermentation and Bioindustrial Biochemistry
17. Food and Feed Chemistry
18. Animal Nutrition
19. Fertilizers, Soils, and Plant Nutrition
20. History, Education, and Documentation

Organic Chemistry Sections

21. General Organic Chemistry
22. Physical Organic Chemistry
23. Aliphatic Compounds
24. Alicyclic Compounds

25. Benzene, Its Derivatives, and Condensed Benzenoid Compounds
26. Biomolecules and Their Synthetic Analogs
27. Heterocyclic Compounds (One Hetero Atom)
28. Heterocyclic Compounds (More Than One Hetero Atom)
29. Organometallic and Organometalloidal Compounds
30. Terpenes and Terpenoids
31. Alkaloids
32. Steroids
33. Carbohydrates
34. Amino Acids, Peptides, and Proteins

Macromolecular Chemistry Sections

35. Chemistry of Synthetic High Polymers
36. Physical Properties of Synthetic High Polymers
37. Plastics Manufacture and Processing
38. Plastics Fabrication and Uses
39. Synthetic Elastomers and Natural Rubber
40. Textiles and Fibers
41. Dyes, Organic Pigments, Fluorescent Brighteners, and Photographic Sensitizers
42. Coatings, Inks, and Related Products
43. Cellulose, Lignin, Paper, and Other Wood Products
44. Industrial Carbohydrates
45. Industrial Organic Chemicals, Leather, Fats, and Waxes
46. Surface-Active Agents and Detergents

Applied Chemistry and Chemical Engineering Sections

47. Apparatus and Plant Equipment
48. Unit Operations and Processes
49. Industrial Inorganic Chemicals
50. Propellants and Explosives
51. Fossil Fuels, Derivatives, and Related Products
52. Electrochemical, Radiational, and Thermal Energy Technology
53. Mineralogical and Geological Chemistry
54. Extractive Metallurgy
55. Ferrous Metals and Alloys
56. Nonferrous Metals and Alloys

57. Ceramics
58. Cement, Concrete, and Related Building Materials
59. Air Pollution and Industrial Hygiene
60. Waste Treatment and Disposal
61. Water
62. Essential Oils and Cosmetics
63. Pharmaceuticals
64. Pharmaceutical Analysis

Physical, Inorganic, and Analytical Chemistry Sections

65. General Physical Chemistry
66. Surface Chemistry and Colloids
67. Catalysis, Reaction Kinetics, and Inorganic
       Reaction Mechanisms
68. Phase Equilibriums, Chemical Equilibriums, and
       Solutions
69. Thermodynamics, Thermochemistry, and Thermal
       Properties
70. Nuclear Phenomena
71. Nuclear Technology
72. Electrochemistry
73. Optical, Electron, and Mass Spectroscopy and
       Other Related Properties
74. Radiation Chemistry, Photochemistry, and
       Photographic and Other Reprographic Processes
75. Crystallography and Liquid Crystals
76. Electric Phenomena
77. Magnetic Phenomena
78. Inorganic Chemicals and Reactions
79. Inorganic Analytical Chemistry
80. Organic Analytical Chemistry

B.4.e.2.   INDEX TO CHEMICAL ABSTRACTS SECTIONS

A

Agents, Surface-Active 46
Agrochemical Bioregulators 5
Air Pollution and Industrial Hygiene 59
Algal, Microbial, and Fungal Biochemistry 10
Alicyclic Compounds 24
Aliphatic Compounds 23
Alkaloids 31

Alloys and Ferrous Metals 55
Alloys and Nonferrous Metals 56
Amino Acids, Peptides, and Proteins 34
Analysis, Pharmaceutical 64
Analytical Inorganic Chemistry 79
Analytical Organic Chemistry 80
Animal Nutrition 18
Apparatus and Plant Equipment 47

B
Benzene, Its Derivatives, and Condensed Benzenoid Compounds 25
Biochemical Genetics 3
Biochemical Methods 9
Biochemistry, Fermentation and Bioindustrial 16
Biochemistry, General 6
Biochemistry, Mammalian 13
Biochemistry, Mammalian Pathological 14
Biochemistry, Microbial, Algal, and Fungal 10
Biochemistry, Nonmammalian 12
Biochemistry, Plant 11
Biochemistry, Radiation 8
Bioindustrial and Fermentation Biochemistry 16
Biomolecules and Their Synthetic Analogs 26
Bioregulators, Agrochemical 5
Brighteners, Fluorescent 41
Building Materials 58

C
Carbohydrates 33
Carbohydrates, Industrial 44
Catalysis, Reaction Kinetics, and Inorganic Reaction Mechanisms 67
Cellulose 43
Cement, Concrete, and Related Building Materials 58
Ceramics 57
Chemical Equilibrium and Solutions 68
Chemicals, Industrial Inorganic 49
Chemicals, Industrial Organic 45
Chemistry, Food and Feed 17
Coatings, Inks, and Related Products 42
Colloids 66
Concrete and Related Building Materials 58
Cosmetics 62

Crystallography and Liquid Crystals 75
Crystals, Liquid 75

D
Detergents 46
Documentation 20
Dyes 41

E
Education and Documentation 20
Elastomers and Natural Rubber 39
Electric Phenomena 76
Electrochemical, Radiational, and Thermal Energy Technology 52
Electrochemistry 72
Electron, and Mass Spectroscopy and Other Related Properties 73
Energy Technology, Thermal 52
Enzymes 7
Equilibriums, Chemical 68
Equilibriums, Phase 68
Equipment, Plant 47
Essential Oils and Cosmetics 62
Explosives 50
Extractive Metallurgy 54

F
Fats 45
Feed and Food Chemistry 17
Fermentation and Bioindustrial Biochemistry 16
Ferrous Metals and Alloys 55
Fertilizers, Soils, and Plant Nutrition 19
Fibers and Textiles 40
Fluorescent Brighteners, 41
Food and Feed Chemistry 17
Fossil Fuels, Derivatives, and Related Products 51
Fuels, Fossil, and Related Products 51
Fungal, Microbial, and Algal Biochemistry 10

G
General Biochemistry 6
Genetics, Biochemical 3
Geological Chemistry 53

H
Heterocyclic Compounds (More Than One Hetero Atom) 28
Heterocyclic Compounds (One Hetero Atom) 27
High Polymers, Chemistry of 35
High Polymers, Physical Properties of 36
History, Education, and Documentation 20
Hormones, Mammalian 2
Hygiene, Industrial 59

I
Immunochemistry 15
Industrial Carbohydrates 44
Industrial Hygiene and Air Pollution 59
Industrial Inorganic Chemicals 49
Industrial Organic Chemicals 45
Inks, Coatings, and Related Products 42
Inorganic Analytical Chemistry 79
Inorganic Chemicals, Industrial 49
Inorganic Chemicals and Reactions 78
Inorganic Reaction Mechanisms 67
Inorganic Reactions 78

K
Kinetics, Reaction 67

L
Leather 45
Lignin 43
Liquid Crystals 75

M
Magnetic Phenomena 77
Mammalian Biochemistry 13
Mammalian Hormones 2
Mammalian Pathological Biochemistry 14
Mass Spectroscopy and Other Related Properties 73
Mechanisms, Inorganic Reaction 67
Metallurgy, Extractive 54
Metals, Ferrous and Alloys 55
Metals, Nonferrous and Alloys 56
Methods, Biochemical 9
Microbial, Algal, and Fungal Biochemistry 10
Mineralogical and Geological Chemistry 53

N

Natural Rubber and Synthetic Elastomers 39
Nonferrous Metals and Alloys 56
Nonmammalian Biochemistry 12
Nuclear Phenomena 70
Nuclear Technology 71
Nutrition, Animal 18
Nutrition, Fertilizers, Soils, and Plant 19

O

Oils, Essential 62
Operations and Processes 48
Optical, Electron, and Mass Spectroscopy and Related Properties 73
Organic Analytical Chemistry 80
Organic Chemicals, Industrial 45
Organic Chemistry, General 21
Organic Chemistry, Physical 22
Organic Pigments 41
Organometallic and Organometalloidal Compounds 29

P

Paper 43
Pathological Biochemistry, Mammalian 14
Peptides and Proteins 34
Pharmaceutical Analysis 64
Pharmaceuticals 63
Pharmacology 1
Phase Equilibriums, Chemical Equilibriums, and Solutions 68
Photochemistry 74
Photographic Sensitizers 41
Photographic and Other Reprographic Processes 74
Physical Chemistry, General 65
Physical Organic Chemistry 22
Pigments, Organic 41
Plant Biochemistry 11
Plant Equipment 47
Plant Nutrition 19
Plastics Manufacture and Processing 37
Plastics Fabrication and Uses 38
Pollution, Air 59
Polymers, Chemistry of Synthetic High 35
Polymers, Physical Properties of Synthetic High 36
Processes 48

Propellants and Explosives 50
Proteins 34

R
Radiation Biochemistry 8
Radiation Chemistry 74
Radiational and Thermal Energy Technology 52
Reaction Kinetics and Inorganic Reaction Mechanisms 67
Reaction Mechanisms, Inorganic 67
Rubber and Synthetic Elastomers 39

S
Sensitizers, Photographic 41
Soils and Plant Nutrition 19
Solutions 68
Spectroscopy, Electron, and Mass and Other Related Properties 73
Spectroscopy, Optical, Electron, and Mass and Related Properties 73
Spectroscopy, Mass and Other Related Properties 73
Steroids 32
Surface Chemistry and Colloids 66
Surface-Active Agents and Detergents 46
Synthetic Elastomers and Natural Rubber 39
Synthetic High Polymers, Chemistry of 35
Synthetic High Polymers, Physical Properties of 36

T
Terpenes and Terpenoids 30
Textiles and Fibers 40
Thermal Energy Technology 52
Thermal Properties 69
Thermochemistry, and Thermal Properties 69
Thermodynamics, Thermochemistry, and Thermal Properties 69
Toxicology 4

U
Unit Operations and Processes 48

W
Waste Disposal and Treatment 60
Waste Treatment and Disposal 60
Water 61
Waxes 45
Wood Products 43

B.5.

The Chemist's Companion:

QD65.G64

- **conversion and mathematical tables**
- **experimental techniques**
- **spectroscopic instrumentation**
- **physical and chemical properties**

B.5.a.   OVERVIEW

The treatise covers properties of chemical compounds, analytical techniques, photochemistry, kinetics and thermodynamics, various experimental and spectroscopic techniques, and conversion and mathematical information.

EASY ACCESS

— Consult the Subject Index in the back of the book.
— The Index to the Table of Contents is located in the organization topic of this section.

INDEXING

The **Subject Index** lists all compounds, properties, general subjects, etc. with the page number where the entry can be found.

The **Suppliers Index** links all referenced suppliers and manufacturers with the page on which their address appears.

ORGANIZATION

There is extensive cross-referencing of chemicals throughout the text, and blank spaces are left in tables for unreliable or unavailable data. **The Chemist's Companion** is organized into nine separate, individualized sections. The titles of the sections and subsections are listed alphabetically below using the following abbreviations:

CHROM — CHROMATOGRAPH
EXPT — EXPERIMENTAL TECHNIQUES
KINETICS — KINETICS AND ENERGETICS
MATH — MATHEMATICAL AND NUMERICAL INFORMATION
MISC — MISCELLANEOUS
PHOTO — PHOTOCHEMISTRY
ATOMS — PROPERTIES OF ATOMS AND BONDS
MOL — PROPERTIES OF MOLECULAR SYSTEMS
SPEC — SPECTROSCOPY

B.5.b.   INDEX TO THE TABLE OF CONTENTS OF THE CHEMIST'S COMPANION

The Chemist's Companion: A Handbook of Practical Data, Techniques, and References (1973), Gordon AJ, Ford RA (ed.). Reprinted by permission of John Wiley & Sons, Inc.

A

Absorption and Emission Spectra: UV and Vis SPEC, Part V.
Acids and Bases MOL, Part IX.
Actinometry, Chemical: Quantum Yield PHOTO, Part IV.
Activation Parameters and Kinetics of Reactions KINETICS, Part I.
Addresses of Publishers that Deal with Chemistry MISC, Part III.
Adsorption Chromatography CHROM, Part II.
Amino Acids, Structure and Properties MOL, Part VI.
Analytical Services and Combustion Microanalysis MISC, Part IV.
Angles, Bond and Hybridization ATOMS, Part V.
Applications and Properties of Liquid Crystals MOL, Part VII.
Approved International Units and General Constants MATH, Part I.
Aromaticity ATOMS, Part X.
Atomic and Molecular Models MISC, Part II.
Atomic Radii, van der Waals ATOMS, Part IV.
Automated Liquid Chromatography CHROM, Part VII.
Azeotropic Data MOL, Part II.

B

Barriers, Torsion and Inversion ATOMS, Part VIII.
Bases and Acids MOL, Part IX.
Basic Definitions and Types of Chromatography CHROM, Part I.
Baths and Solvents for Heating and Cooling EXPT, Part XI.
Bibliography of Chemistry Reference Sources MISC, Part I.
Bibliography of Spectral Data Compilations SPEC, Part XI.
Boiling Point-Pressure Relationships MOL, Part III.
Bond and Group Dipole Moments ATOMS, Part IX.

Bond Angles and Hybridization ATOMS, Part V.
Bond Lengths ATOMS, Part III.
Bond Strengths ATOMS, Part VI.

C

Character Tables for Common Symmetry Groups MATH, Part VI.
Chemical Actinometry: Quantum Yield PHOTO, Part IV.
Chemical Methods for Deoxygenating Gases and Liquids EXPT, Part I.
Chemical Methods for Detecting Specific Gases EXPT, Part VI.
Chemicals, Hazards MISC, Part V.
Chemistry Publishers, Addresses MISC, Part III.
Chemistry Reference Sources: A Bibliography MISC, Part I.
Chromatography References CHROM, Part XI.
CHROMATOGRAPHY SECTION
Chromatography Supply Directory CHROM, Part X.
Chromatography Types and Basic Definitions CHROM, Part I.
Cleaning Solutions for Glassware EXPT, Part II.
Column and Thin Layer Partition Chromatography CHROM, Part IV.
Combustion Microanalysis and Analytical Services MISC, Part IV.
Common Solvents for Crystallization EXPT, Part VIII.
Common Solvent Purification EXPT, Part III.
Composition-Free Energy Chart KINETICS, Part IV.
Computer Programs MATH, Part VII.
Conformational Free Energy Values KINETICS, Part III.
Constants and International Units System MATH, Part I.
Conversion Factors MATH, Part II.
Conversion Table, Wavelength-Wavenumber MATH, Part III.
Cooling and Heating Solvents and Baths EXPT, Part XI.
Crystallization Solvents EXPT, Part VIII.
Custom Analytical Services and Combustion Analysis MISC, Part IV.

D

Data, Statistical Treatment of MATH, Part VIII.
Deoxygenating Gases and Liquids by Chemical Methods EXPT, Part I.
Detecting Specific Gases by Chemical Methods EXPT, Part VI.
Detection of Peroxides and Their Removal EXPT, Part IV.
Determination of Molecular Weight EXPT, Part XII.
Dipole Moments, Bond and Group ATOMS, Part IX.
Directory of Chromatography Supply CHROM, Part X
Dry Gases, Simple Preparations of EXPT, Part VIII.
Drying Agents EXPT, Part X.

E
Effective van der Waals Radii ATOMS, Part IV.
Electromagnetic Spectrum SPEC, Part I.
Electron Spin Resonance Spectroscopy SPEC, Part IX.
Electronic Energy State Diagram PHOTO, Part I.
Electrophoresis CHROM, Part VIII.
Element and Group Weights, Multiples of MATH, Part IV.
Elements, Properties ATOMS, Part I.
Emission and Absorption Spectra: UV and Vis SPEC, Part V.
Empirical Boiling Point-Pressure Relationships MOL, Part III.
Energy-Composition Chart KINETICS, Part IV.
Energy Relationships, Linear KINETICS, Part II.
Energy State Diagram, Electronic PHOTO, Part I.
Energy Transfer: Sensitizers and Quenchers PHOTO, Part II.
Energy Values, Conformational KINETICS, Part III.
Equipment and Light Sources, Photochemistry PHOTO, Part III.
ESR Spectroscopy SPEC, Part IX.
Excited State Energy, Sensitizers and Quenchers PHOTO, Part II.
EXPERIMENTAL TECHNIQUES SECTION
Extraction Solvents for Aqueous Solutions EXPT, Part IX.

F
Force Constants ATOMS, Part VII.
Free Energy-Composition Chart KINETICS, Part IV.
Free Energy Relationships, Linear KINETICS, Part II.
Free Energy Values, Conformational KINETICS, Part III.
Fundamental Chromatography and Definitions CHROM, Part I.
Fused Salt Systems, Properties MOL, Part V.

G
Gas Phase Chromatography CHROM, Part IX.
Gases and Liquids, Methods for Deoxygenating EXPT, Part I.
Gases, Detecting by Simple Chemical Methods EXPT, Part VI.
Gases, Dry, Preparations of EXPT, Part VIII.
Gases, Properties MOL, Part IV.
GC CHROM, Part IX.
Gel Permeation and Gel Filtration Chromatography CHROM, Part VI.
Gel Filtration and Gel Permeation Chromatography CHROM, Part VI.
General Constants and Approved International Units MATH, Part I.
Glassware Cleaning Solutions EXPT, Part II.
Group and Bond Dipole Moments ATOMS, Part IX.
Group Weights and Elements, Multiples of MATH, Part IV.

H
Hazards of Common Chemicals MISC, Part V
Heating and Cooling Solvents and Baths EXPT, Part XI.
Hybridization and Bond Angles ATOMS, Part V.

I
Important Chemistry Reference Sources MISC, Part I.
INDEX, SUBJECT
INDEX, SUPPLIERS
Infrared Spectra SPEC, Part IV.
Inversion and Torsion Barriers ATOMS, Part VIII.
Ion-Exchange Chromatography CHROM, Part V.
Isotopes, Table ATOMS, Part II.

K
Kinetics and Activation Parameters of Reactions KINETICS, Part I.
KINETICS AND ENERGETICS SECTION

L
Laboratory Materials, Properties EXPT, Part I.
LC CHROM, Part VII.
Lengths, Bond ATOMS, Part III.
Light Sources and Equipment, Photochemistry PHOTO, Part III.
Linear Free Energy Relationships KINETICS, Part II.
Liquid Chromatography CHROM, Part VII.
Liquid Crystals, Properties and Applications MOL, Part VII.
Liquid and Gases, Chemical Methods for Deoxygenating EXPT, Part I.
Liquids and Solvents, Properties MOL, Part I.

M
Mass Spectrometry SPEC, Part VII.
Materials for Spectroscopy and Photochemistry SPEC, Part III.
MATHEMATICAL AND NUMERICAL INFORMATION SECTION
Mathematical Constants and International Units MATH, Part I.
Media and Solvents for Spectral Measurements SPEC, Part II.
Methods for Deoxygenating Gases and Liquids EXPT, Part I.
Methods for Detecting Specific Gases EXPT, Part VI.
Microanalysis and Custom Analytical Services MISC, Part IV.
MISCELLANEOUS SECTION
Models, Atomic and Molecular MISC, Part II.
Molecular and Atomic Models MISC, Part II.
Molecular Symmetry: Definitions and Systems MATH, Part V.

Molecular Weight Determination EXPT, Part XII.
Multiples of Element and Group Weights MATH, Part IV.

N
Naturally Occurring α-Amino Acids MOL, Part VI.
NMR Spectroscopy SPEC, Part VIII.
Nuclear Magnetic Resonance Spectroscopy SPEC, Part VIII.
Nuclear Quadrupole Resonance Spectroscopy SPEC, Part X.

O
Optical Activity and Optical Rotation SPEC, Part III.
Optical Materials for SpectA and Photochemistry SPEC, Part III.
Optical Rotation and Optical Activity SPEC, Part III.

P
Paper Chromatography CHROM, Part III.
Peroxides, Detection and Removal EXPT, Part IV.
Photochemical References PHOTO, Part VI.
Photochemical Suppliers PHOTO, Part V.
Photochemistry and Spectroscopy, Optical Materials SPEC, Part III.
Photochemistry Light Sources and Equipment PHOTO, Part III.
Photochemistry, Quenchers and Sensitizers PHOTO, Part II.
PHOTOCHEMISTRY SECTION
Preparations of Some Dry Gases EXPT, Part VIII.
Pressure-Boiling Point Relationships MOL, Part III.
Programs, Computer MATH, Part VII.
Properties and Applications of Liquid Crystals MOL, Part VII.
Properties and Structure of α-Amino Acids MOL, Part VI.
PROPERTIES OF ATOMS AND BONDS SECTION
Properties of Laboratory Materials EXPT, Part I.
PROPERTIES OF MOLECULAR SYSTEMS SECTION
Properties of Representative Fused Salt Systems MOL, Part V.
Properties of Selected Gases MOL, Part IV.
Properties of Solvents and Common Liquids MOL, Part I.
Properties of the Elements ATOMS, Part I.
Prototropic Tautomerism MOL, Part VIII.
Publishers that Deal with Chemistry, Addresses of MISC, Part III.
Purification of Common Solvents EXPT, Part III.

Q
Quantum Yield, Chemical Actinometry PHOTO, Part IV.
Quenchers and Sensitizers, Photochemistry PHOTO, Part II.

R

Radii, van der Waals ATOMS, Part IV.
Raman Scattering Spectra SPEC, Part IV.
Reactions Kinetics and Activation Parameters KINETICS, Part I.
Recrystallization Solvents EXPT, Part VIII.
References, Chromatography CHROM, Part XI.
References, Photochemical PHOTO, Part VI.
Removal of Peroxides and Their Detection EXPT, Part IV.

S

Salt Systems, Fused, Properties MOL, Part V.
Selected Bond Lengths ATOMS, Part III.
Selected Bond Strengths ATOMS, Part VI.
Sensitizers and Quenchers, Photochemistry PHOTO, Part II.
Simple Chemical Methods for Detecting Gases EXPT, Part VI.
Simple Preparations of Some Dry Gases EXPT, Part VIII.
Solutions for Cleaning Standard Glassware EXPT, Part II.
Solvents and Baths for Heating and Cooling EXPT, Part XI.
Solvents and Common Liquids, Properties MOL, Part I.
Solvents and Other Media for Spectral Measurements SPEC, Part II.
Solvents for Crystallization EXPT, Part VIII.
Solvents for Extraction of Aqueous Solutions EXPT, Part IX.
Solvents, Purification of EXPT, Part III.
Spectral Data Compilations, Bibliography of SPEC, Part XI.
Spectral Measurements, Solvents and Other Media SPEC, Part II.
Spectroscopy and Photochemistry, Optical Materials SPEC, Part III.
SPECTROSCOPY SECTION
Spectrum, Electromagnetic SPEC, Part I.
Standard Glassware Cleaning Solutions EXPT, Part II.
State Diagram, Electronic Energy PHOTO, Part I.
Statistical Treatment of Data MATH, Part VIII.
Strengths, Bond ATOMS, Part VI.
Structure and Properties of α-Amino Acids MOL, Part VI.
SUBJECT INDEX
SUPPLIERS INDEX
Suppliers, Photochemical PHOTO, Part V.
Symmetry Groups, Tables for MATH, Part VI.
Symmetry, Molecular: Definitions and Systems MATH, Part V.

T

Table, Conversion, Wavelength-Wavenumber MATH, Part III.
Table of Isotopes ATOMS, Part II.
Tautomerism, Prototropic MOL, Part VIII.

Thin Layer and Column Chromatography CHROM, Part IV.
Torsion and Inversion Barriers ATOMS, Part VIII.

U
Units System and General Constants MATH, Part I.
Useful Conversion Factors MATH, Part II.
UV and Vis Spectra SPEC, Part V.

V
Vapor Phase Chromatography CHROM, Part IX.
Vibration Spectra SPEC, Part IV.
Vis and UV Spectra, SPEC, Part V.

W
Wavelength-Wavenumber Conversion Table MATH, Part III.
Wavenumber-Wavelength Conversion Table MATH, Part III.

B.6.

Compendium of Organic Synthetic Methods:

QD262.H31

- **functional group transformations**
- **protection of functional groups**
- **preparation of difunctional groups**
- **reviews**

The reactions are classified on the basis of the functional group in the starting material and of the product with no reference to mechanism.

EASY ACCESS

— Consult the appropriate index in the front of the most recent volume to locate the Section (in bold type) containing the desired transformation. The Section number can be cross-referenced between volumes.
— The Section numbers for monofunctional and difunctional transformations can be obtained from the tables at the end of B.6.

INDEXING

The **INDEX, Monofunctional Compounds** is a table located in the front of each volume. The starting functional group is listed along the left vertical axis with the product along the top horizontal. The intersection of the two functional groups of interest gives the Section (in bold type) and page where the transformation can be found. Protecting groups are listed on the right-hand side of the table.

The **INDEX, Difunctional Compounds** (Volume 2 and subsequent volumes) follows the monofunctional compound index in the front of each volume. This table gives the Section (in bold type) and page corresponding to each difunctional product.

The **Author Index** (Volume 6 and subsequent volumes) is located in the back of the volume and is an alphabetical listing of all authors referenced in that volume.

ORGANIZATION

Volume 1 contains only functional group transformations and the protection of carboxylic acids, alcohols, phenols, aldehydes, amines, and ketones. The chapters are organized according to the functional group being prepared and are sectioned into those reactions that have the same starting groups. Chain lengthening processes are listed before degradations. The miscellaneous section of each chapter contains reactions that remove one functional group from a difunctional compound or prepare functional groups from groups not listed in the index. Reviews are listed with the appropriate functional group transformation.

Volume 2 and subsequent volumes update the functional group transformation chapters and include the preparation of difunctional groups and expand the protection of functional groups to include acetylenes, amides, esters, and olefins. Reactions in the difunctional group chapter are divided into sections on the basis of the two functional groups in the product.

EXAMPLE ENTRY

Section 145 <u>Halides from Halides and Sulfonates</u>

Me$_3$SiI, 72%

JOC, 46, 3727 (1981)

Compendium of Organic Synthetic Methods, M. B. Smith, Copyright 1992. Reprinted by permission of John Wiley & Sons, Inc.

INDEX, MONOFUNCTIONAL COMPOUNDS

Sections ONLY

PREPARATION OF →

FROM ↓

FROM ＼ PREPARATION OF	Acetylenes	Carboxylic acids, acidhalides, anhydrides	Alcohols, phenols	Aldehydes	Alkyls, methylenes, aryls	Amides	Amines	Esters	Ethers, epoxides	Halides, sulfonates, sulfates	Hydrides (RH)	Ketones	Nitriles	Olefins	Oxides
Acetylenes	1	16	31	46	61	76	91	106	121	136		166	181	196	
Carboxylic acids acid halides, anhydrides	2	17	32	47	62	77	92	107	122	137	152	167	182	197	
Alcohols, phenols	3	18	33	48	63	78	93	108	123	138	153	168	183	198	213
Aldehydes	4	19	34	49	64	79	94	109	124	139	154	169	184	199	
Alkyls, methylenes, aryls		20	35	50	65	80	95	110	125	140	155	170	185	200	215
Amides		21	36	51	66	81	96	111		141	156	171	186	201	
Amines	7	22	37	52	67	82	97	112	127	142	157	172	187	202	217
Esters	8	23	38	53	68	83	98	113	128	143	158	173	188	203	
Ethers, epoxides		24	39	54	69	84	99	114	129	144	159	174	189	204	219
Halides, sulfonates, sulfates	10	25	40	55	70	85	100	115	130	145	160	175	190	205	
Hydrides (RH)		26	41	56	71	86	101	116	131	146	161	176	191	206	221
Ketones	12	27	42	57	72	87	102	117	132	147	162	177	192	207	
Nitriles		28	43	58	73	88	103	118			163	178	193	208	
Olefins	14	29	44	59	74	89	104	119	134	149	164	179	194	209	
Miscellaneous compounds	15	30	45	60	75	90	105	120	135	150	165	180	195	210	225

PROTECTION

Acetylenes 15A
Carboxylic acids 30A
Alcohols, phenols 45A
Aldehydes 60A
Amides 90A
Esters 105A
Ketones 180A
Olefins 210A

INDEX, DIFUNCTIONAL COMPOUNDS

Sections ONLY

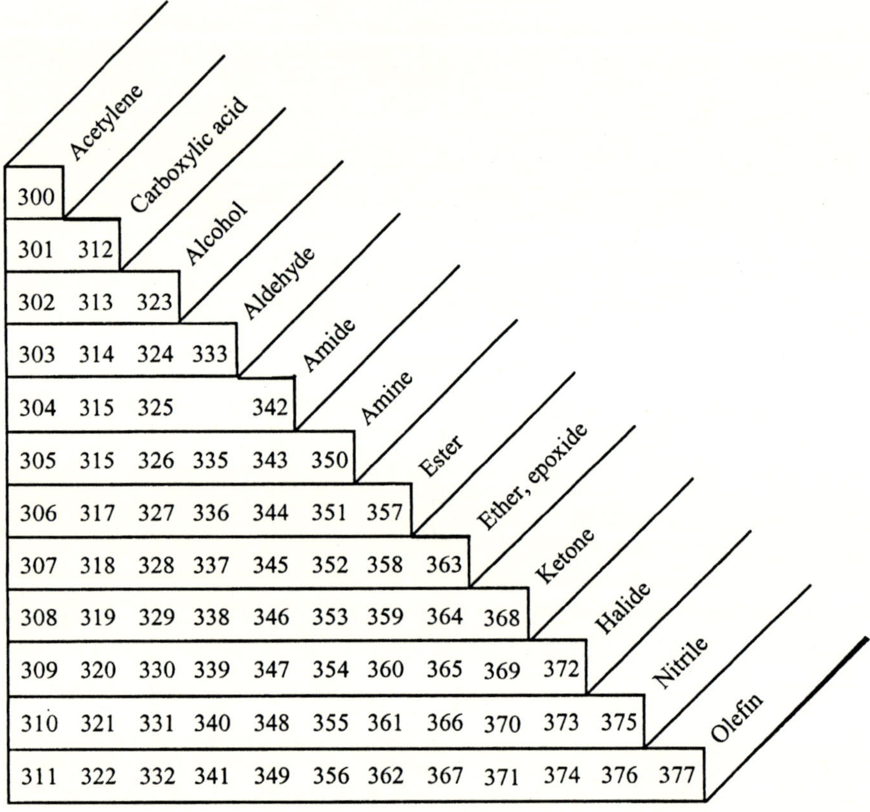

Blanks in the table correspond to sections for which no examples were found in the literature.

Compendium of Organic Synthetic Methods, M. B. Smith, Copyright 1992. Reprinted by permission of John Wiley & Sons, Inc.

B.7.

Comprehensive Organic Chemistry:

QD245.C65

- **preparation and reactions of *classes* of compounds**
- **rearrangements**

B.7.a. OVERVIEW

The treatise covers the preparation, properties, and reactions of functional groups in organic compounds. Modern synthetic methods and name reactions are included. Extensive literature references accompany each review.

EASY ACCESS

— For chemical substances, determine the correct name and reference in the Formula Index of Volume 6.
— Check the Index of the major sections under Organization.

INDEXING

Volume 6 contains the Author Index, Formula Index, Reaction Index, Reagent Index, and Subject Index. The **Author Index** lists the author's name followed in bold numbers by the volume, ordinary numbers to the pages, and italic numbers to the references. If a reference is cited on a particular page more than once, this is indicated by a superscript number for the number of times it is cited. When the reference is cited as a page, table, or scheme footnote, the letter "f" is placed after the page number.

The **Formula Index** lists the molecular formula of carbon-containing substances that are significant for the preparation or reactions of the functional groups. Formulas are arranged by citing carbons in ascending order, then hydrogens, and finally the remaining elements in alphabetical order.

The **Reaction Index** lists named reactions and specific types of transformations. This section also lists reagents or combinations of reagents needed to carry out the reaction with pertinent literature references.

The **Reagent Index** lists any organic or inorganic substance used in organic synthesis along with those compounds that can be transformed by the reagent. Literature references describing its use are included in this index.

The **Subject Index** contains entries to classes of organic functional groups, individual compounds and their derivatives, separation techniques, and spectroscopic methods mentioned in the text. This index does *not* include named reactions, uses of organic or inorganic compounds, and types of reactions.

ORGANIZATION

Comprehensive Organic Chemistry is a six-volume set. The first five volumes cover the synthesis and reactions of organic compounds. Those topics covered in each volume are outlined on its spine. The sixth volume is the general index. The major sections are listed alphabetically below.

B.7.b.   INDEX TO THE TABLE OF CONTENTS OF COMPREHENSIVE ORGANIC CHEMISTRY

Reprinted from *Comprehensive Organic Chemistry* (1979), D. Barton and W. D. Ollis (eds.) with permission from Elsevier Science.

A
Alcohols and Related Compounds, Vol. 1, Part 4
Aldehydes, Vol. 1, Part 5
Amines and Related Compounds, Vol. 2, Part 6
Amino Acids, Vol. 5, Part 23
Author Index, Vol. 6
Azines, Vol. 4, Part 16
Azoles, Vol. 4, Part 17

B
Bio-Organic Chemistry, Vol. 5, Part 28
Biological Compounds, Vol. 5
Biological Chemistry, Vol. 5, Part 21
Biosynthesis, Vol. 5, Part 30
Biosynthetic Pathways from Acetate, Vol. 5, Part 29
Boron Organic Compounds, Vol. 3, Part 14

C
Carbohydrate Chemistry, Vol. 5, Part 26
Carbon Compounds, Vol. 1
Carboxylic Acids and Related Compounds, Vol. 2, Part 9
Carboxylic Acids, Vol. 2

E
Enzyme Catalysis, Vol. 5, Part 24
Ethers and Related Compounds, Vol. 1, Part 4

F
Formula Index, Vol. 6

H
Halogen Compounds, Vol. 1, Part 3
Heterocyclic Compounds, Vol. 4
Heterocyclics with Heteroatoms (not N, O, or S) Systems, Vol. 4, Part 19
Hydrocarbons, Vol. 1, Part 2

I
Imines, Vol. 2, Part 8
Indexes, Vol. 6
Isocyanides, Vol. 2, Part 8

K
Ketones, Vol. 1, Part 5

L
Lipid Chemistry, Vol. 5, Part 25

M
Macromolecules, Vol. 5, Part 27
Mixed Heteroatoms Heterocyclic Systems, Vol. 4, Part 20

N
Nitriles, Vol. 2, Part 8
Nitro Compounds, Vol. 2, Part 7
Nitrogen Compounds, Vol. 2
Nitrogen Heterocyclic Systems, Vol. 4, Part 17
Nitrones, Vol. 2, Part 8
Nitroso Compounds, Vol. 2, Part 7
Nomenclature, Vol. 1, Part 1
Nucleic Acids, Vol. 5, Part 22

O
Organometallic Compounds, Vol. 3, Part 15
Oxygen Compounds, Vol. 1
Oxygen Heterocyclic Systems, Vol. 4, Part 18

P
Peptides, Vol. 5, Part 23
Phenols and Related Compounds, Vol. 1, Part 4
Phosphorus Compounds, Vol. 2, Part 10
Proteins, Vol. 5, Parts 23 and 24
Proteins, Functional, Vol. 5, Part 24

R
Reaction Index, Vol. 6
Reagent Index, Vol. 6

S
Selenium Organic Compounds, Vol. 3, Part 12
Silicon Organic Compounds, Vol. 3, Part 13
Stereochemistry, Vol. 1, Part 1
Subject Index, Vol. 6
Sulfur Heterocyclic Systems, Vol. 4, Part 19
Sulfur Organic Compounds, Vol. 3, Part 11

T
Tellurium Organic Compounds, Vol. 3, Part 12

B.8.

Comprehensive Organometallic Chemistry:

QD411.C652

- **properties and synthesis of *classes* of compounds**
- **catalysis**
- **organometallic reviews**
- **X-ray structures**

B.8.a. OVERVIEW

The treatise covers the preparation, properties, and reactions of organometallic compounds. A review of the chemistry of the elements is also included. Extensive literature references accompany each review.

EASY ACCESS

— For chemical substances, determine the correct name and reference in the Formula Index of Volume 9.

INDEXING

Volume 9 contains the Author Index, Formula Index, Subject Index, Index of Review Articles and Specialist Texts on Organometallic Chemistry, and Index of Structures Determined by Diffraction Methods. The **Author Index** lists the author's name followed in bold numbers by the volume; ordinary numbers to the chapter bibliography; superscript numbers to the reference number in the chapter bibliography; the pages where the reference is cited are in parentheses. When a reference is cited in a table, the letter "T" appears after the page number. When the reference is a footnote to a table, the letter "f" is placed after the page number and a superscript number is used to indicate the reference number within the table.

The **Formula Index** lists the molecular formula of every substance containing a metal–carbon bond and considered important in the context of a particular topic under discussion. Formulas are arranged by citing metal(s) first, then carbons in ascending order, hydrogens, and finally the remaining elements in alphabetical order. Compounds containing two (or more) different metals are listed under each metal.

The **Subject Index** contains entries to classes of organometallic compounds, individual compounds, types of reactions, separation techniques, and spectroscopic methods, and organic compounds when their synthesis or use involves organometallic compounds.

The **Index of Review Articles and Specialist Texts on Organometallic Chemistry** provides a comprehensive coverage of review literature from 1970 to 1981. All references are collected in a numbered, titled, unclassified Main Index. A series of Access Tables provide reference information on a specific topic. A number in parentheses, accompanying the index reference number, indicates the chapter of the book referenced.

The **Index of Structures Determined by Diffraction Methods** lists all compounds containing at least one carbon–metal bond whose structure has been determined by X-rays (X), electrons (E), or neutrons (N) diffraction methods from 1927 to 1981. Compounds are arranged in alphabetical order by citing all metal(s) first, then carbons in ascending order, hydrogens, and finally the remaining elements. An asterisk (*) following a reference indicates that the entry is a preliminary account and a full report has appeared.

ORGANIZATION

Comprehensive Organometallic Chemistry is a nine-volume set. The first eight volumes cover the synthesis, reactions, and structures of organometallic compounds. Volume 9 is the general index. The Table of Contents is indexed below.

B.8.b. INDEX TO THE TABLE OF CONTENTS OF COMPREHENSIVE ORGANOMETALLIC CHEMISTRY

Reprinted from *Comprehensive Organometallic Chemistry* (1995), E. W. Abel (ed.) with permission from Elsevier Science.

A

Actinides, Vol. 3, Part 21

Addition of Hydrogen and HCN to Double and Triple Bonds, Vol. 8, Part 51

Alkali and Alkaline Earth Metal Compounds in Organic Synthesis, Vol. 7, Part 44

Alkali Metals, Vol. 1, Part 2

Alkaline and Alkali Earth Metal Compounds in Organic Synthesis, Vol. 7, Part 44

Alkene and Alkyne Metathesis Reactions, Vol. 8, Part 54

Alkene Oligomerization, Cooligomerization and Telomerization Reactions, Vol. 8, Part 52

Alkyne and Alkene Metathesis Reactions, Vol. 8, Part 54
Alkyne Oligomerization, Cooligomerization and Telomerization Reactions, Vol. 8, Part 52
Aluminum, Vol. 1, Part 6
Aluminum Compounds in Organic Synthesis, Vol. 7, Part 46
Aluminum Compounds with Bonds to a Transition Metal, Vol. 6, Part 41
Antimony, Vol. 2, Part 13
Arene and Cyclopentadienyl Metal-Coordinated Ligands in Org Chem, Vol. 8, Part 59
Arsenic, Vol. 2, Part 13
Asymmetric Synthesis Using Organometallic Catalysts, Vol. 8, Part 53
Author Index, Vol. 9

B
Barium, Vol. 1, Part 4
Beryllium, Vol. 1, Part 3
Bismuth, Vol. 2, Part 13
Bonding and Structural Relationships among Organometallics, Vol. 1, Part 1
Bonds Between a Transition Metal and B, Al, Ga, In or Tl, Vol. 6, Part 41
Bonds Between a Transition Metal and Hg, Cd, Zn or Mg, Vol. 6, Part 42
Bonds Between a Transition Metal and Si, Ge, Sn or Pb, Vol. 6, Part 43
Bonding of Unsaturated Organic Molecules to Transition Metals, Vol. 3, Part 19
Boron, Vol. 1, Part 5
Boron Compounds in Organic Synthesis, Vol. 7, Part 45
Boron Compounds with Bonds to a Transition Metal, Vol. 6, Part 41

C
Cadmium, Vol. 2, Part 16
Cadmium Compounds in Organic Synthesis, Vol. 7, Part 49
Cadmium Compounds with Bonds to a Transition Metal, Vol. 6, Part 42
Calcium, Vol. 1, Part 4
Carbon Monoxide and Carbon Dioxide in Organic Synthesis, Vol. 8, Part 50
Catalysis and Compounds, Organopalladium, in Organic Synthesis, Vol. 8, Part 57
Catalysts, Organometallic, for Asymmetric Synthesis, Vol. 8, Part 53
Catalysts, Polymer Supported, Vol. 8, Part 55
Chromium, Vol. 3, Part 26

Cobalt, Vol. 5, Part 34

Compounds of Aluminum in Organic Synthesis, Vol. 7, Part 46

Compounds of the Alkali and Alkaline Earth Metals in Synthesis, Vol. 7, Part 44

Compounds of Thallium in Organic Synthesis, Vol. 7, Part 47

Compounds of Zn, Cd, Hg, Cu, Ag, and Au in Organic Synthesis, Vol. 7, Part 49

Compounds with Bonds Between a Transition Metal and B, Al, Ga, In, or Tl, Vol. 6, Part 41

Compounds with Bonds Between a Transition Metal and Hg, Cd, Zn, or Mg, Vol. 6, Part 42

Compounds with Bonds Between a Transition Metal and Si, Ge, Sn, or Pb, Vol. 6, Part 43

Compounds with Heteronuclear Bonds between Transition Metals, Vol. 6, Part 40

Cooligomerization, Oligomerization, and Telomerization Reactions, Vol. 8, Part 52

Copper, Vol. 2, Part 14

Copper Compounds in Organic Synthesis, Vol. 7, Part 49

Cyclopentadienyl and Arene Metal-coordinated Ligands in Organic Chemistry, Vol. 8, Part 59

D

Diffraction Methods, Index of Structures Determined by, Vol. 9

Dinitrogen Reactions Promoted by Transition Metal Compounds, Vol. 8, Part 60

Double and Triple Bonds Addition of Hydrogen and HCN, Vol. 8, Part 51

E

Electron Diffraction Methods, Index of Structures Determined by, Vol. 9

Environmental Aspects of Organometallic Chemistry, Vol. 2, Part 18

F

Formula Index, Vol. 9

G

Gallium, Vol. 1, Part 7

Gallium Compounds with Bonds to a Transition Metal, Vol. 6, Part 41

Germanium, Vol. 2, Part 10

Germanium Compounds with Bonds to a Transition Metal, Vol. 6, Part 43
Gold, Vol. 2, Part 15
Gold Compounds in Organic Synthesis, Vol. 7, Part 49

H
Hafnium, Vol. 3, Part 23
Heteronuclear Bonds Between Transition Metals and Other Compounds, Vol. 6, Part 40
Hydrogen and HCN Addition to Double and Triple Bonds, Vol. 8, Part 51
Hydrogen Cyanide and Hydrogen Addition to Double and Triple Bonds, Vol. 8, Part 51

I
Indexes, Vol. 9
Index of Review Articles and Specialist Texts on Organometallic Chemistry, Vol. 9
Index of Structures Determined by Diffraction Methods, Vol. 9
Indium, Vol. 1, Part 7
Indium Compounds with Bonds to a Transition Metal, Vol. 6, Part 41
Iridium, Vol. 5, Part 36
Iron, Vol. 4, Part 31
Iron Compounds in Stoichiometric Organic Synthesis, Vol. 8, Part 58.

L
Lanthanides, Vol. 3, Part 21
Lead, Vol. 2, Part 12
Lead Compounds with Bonds to a Transition Metal, Vol. 6, Part 43

M
Manganese, Vol. 4, Part 29
Magnesium, Vol. 1, Part 4
Magnesium Compounds with Bonds to a Transition Metal, Vol. 6, Part 42
Mercury, Vol. 2, Part 17
Mercury Compounds in Organic Synthesis, Vol. 7, Part 49
Mercury Compounds with Bonds to a Transition Metal, Vol. 6, Part 42
Metal-Coordinated Cyclopentadienyl and Arene Ligands in Organic Chemistry, Vol. 8, Part 59
Metathesis Reactions of Alkenes and Alkynes, Vol. 8, Part 54
Molybdenum, Vol. 3, Part 27

N

Neutrons Diffraction Methods, Index of Structures Determined by, Vol. 9

Nickel, Vol. 6, Part 37

Nickel Compounds in Organic Synthesis, Vol. 8, Part 56

Niobium, Vol. 3, Part 25

Non-rigidity in Organometallic Compounds, Vol. 3, Part 20

O

Oligomerization, Cooligomerization and Telomerization Reactions, Vol. 8, Part 52

Organic Synthesis with Alkali and Alkaline Earth Metal Compounds in, Vol. 7, Part 44

Organic Synthesis with Carbon Monoxide and Carbon Dioxide, Vol. 8, Part 50

Organic Synthesis with Compounds of Aluminum, Vol. 7, Part 46

Organic Synthesis with Compounds of Thallium, Vol. 7, Part 47

Organic Synthesis with Organoboron Compounds, Vol. 7, Part 45

Organic Synthesis with Organonickel Compounds in, Vol. 8, Part 56

Organic Synthesis with Organopalladium Compounds and Catalysis, Vol. 8, Part 57

Organic Synthesis with Organosilicon Compounds, Vol. 7, Part

Organic Synthesis with Stoichiometric Organoiron Compounds, Vol. 8, Part 58

Organic Synthesis with Zn, Cd, Hg, Cu, Ag, and Au Compounds, Vol. 7, Part 49

Organoboron Compounds in Organic Synthesis, Vol. 7, Part 45

Organoiron Compounds in Stoichiometric Organic Synthesis, Vol. 8, Part 58

Organometallics, Bonding and Structural Relationships, Vol. 1, Part 1

Organometallic Catalysts for Asymmetric Synthesis, Vol. 8, Part 53

Organometallic Chemistry, Environmental Aspects, Vol. 2, Part 18

Organometallic Chemistry, Index of Review Articles and Specialist Texts, Vol. 9

Organometallic Compounds, Non-rigidity, Vol. 3, Part 20

Organonickel Compounds in Organic Synthesis, Vol. 8, Part 56

Organopalladium Compounds in Organic Synthesis and in Catalysis, Vol. 8, Part 57

Organosilicon Compounds in Organic Synthesis, Vol. 7, Part 48

Osmium, Vol. 4, Part 33

P

Palladium, Vol. 6, Part 38

Palladium Compounds and Catalysis in Organic Synthesis, Vol. 8, Part 57

Platinum, Vol. 6, Part 39

Polymer Supported Catalysts, Vol. 8, Part 55

R

Review Articles and Specialist Texts on Organometallic Chemistry, Index of, Vol. 9

Rhenium, Vol. 4, Part 30

Rhodium, Vol. 5, Part 35

Ruthenium, Vol. 4, Part 32

S

Scandium, Vol. 3, Part 21

Silicon, Vol. 2, Part 9

Silicon Compounds with Bonds to a Transition Metal, Vol. 6, Part 43

Silver, Vol. 2, Part 14

Silver Compounds in Organic Synthesis, Vol. 7, Part 49

Specialist Texts and Review Articles on Organometallic Chemistry, Index of, Vol. 9

Strontium, Vol. 1, Part 4

Structural and Bonding Relationships among Organometallics, Vol. 1, Part 1

Structures Determined by Diffraction Methods, Index, Vol. 9

Subject Index, Vol. 9

Supported Catalysts, Polymer, Vol. 8, Part 55

Synthesis, Asymmetric using Organometallic Catalysts, Vol. 8, Part 53

Synthesis with Alkali and Alkaline Earth Metal Compounds in, Vol. 7, Part 44

Synthesis with Carbon Monoxide and Carbon Dioxide, Vol. 8, Part 50

Synthesis with Compounds of Aluminum, Vol. 7, Part 46

Synthesis with Compounds of Thallium, Vol. 7, Part 47

Synthesis with Organoboron Compounds, Vol. 7, Part 45

Synthesis with Organonickel Compounds, Vol. 8, Part 56

Synthesis with Organopalladium Compounds and Catalysis, Vol. 8, Part 57

Synthesis with Organosilicon Compounds, Vol. 7, Part

Synthesis with Stoichiometric Organoiron Compounds, Vol. 8, Part 58

Synthesis with Zn, Cd, Hg, Cu, Ag and Au Compounds, Vol. 7, Part 49

T

Tantalum, Vol. 3, Part 25

Technetium, Vol. 4, Part 30

Telomerization, Oligomerization, Cooligomerization Reactions, Vol. 8, Part 52

Thallium, Vol. 1, Part 8

Thallium Compounds in Organic Synthesis, Vol. 7, Part 47

Thallium Compounds with Bonds to a Transition Metal, Vol. 6, Part 41

Tin Compounds with Bonds to a Transition Metal, Vol. 6, Part 43

Tin, Vol. 2, Part 11

Titanium, Vol. 3, Part 22

Transition Metal, Bonding to Unsaturated Organic Molecules, Vol. 3, Part 19

Transition Metal, Compounds with Bonds between B, Al, Ga, In, or Tl, Vol. 6, Part 41

Transition Metal Compounds with Bonds between Hg, Cd, Zn, or Mg, Vol. 6, Part 42

Transition Metal Compounds with Bonds between Si, Ge, Sn, or Pb, Vol. 6, Part 43

Transition Metal Promoted Reactions of Dinitrogen, Vol. 8, Part

Transition Metal, with Heteronuclear Bonds, Vol. 6, Part 40

Triple and Double Bonds Addition of Hydrogen and HCN, Vol. 8, Part 51

Tungsten, Vol. 3, Part 28

U

Unsaturated Organic Molecules, Bonding to Transition Metals, Vol. 3, Part 19

V

Vanadium, Vol. 3, Part 24

X

X-ray Diffraction Methods, Index of Structures Determined by, Vol. 9

Y

Yttrium, Vol. 3, Part 21

Z

Zinc, Vol. 2, Part 16

Zinc, Compounds in Organic Synthesis, Vol. 7, Part 49

Zinc Compounds with Bonds to a Transition Metal, Vol. 6, Part 42

Zirconium, Vol. 3, Part 23

B.9.

Comprehensive Organic Functional Group Transformations:

QD262.C534

- **introduction of functional groups**
- **interconversion of functional groups**

The treatise covers the preparation, properties, and reactions of functional groups in organic compounds. Modern synthetic methods and name reactions are included. Extensive literature references accompany each review.

EASY ACCESS

— For chemical substances, determine the correct name and reference in the Formula Index of Volume 7.

INDEXING

Volume 7 contains the subject and author indexes.

The **Formula Index** lists the molecular formula of carbon-containing substances that are significant for the preparation or reactions of the functional group under discussion. Formulas are arranged by citing carbons in ascending order, then hydrogens, and finally the remaining elements in alphabetical order.

The **Author Index** links the name of the authors with the volume and page where their work is referenced.

ORGANIZATION

Comprehensive Organic Chemistry is a seven-volume set. The first six volumes cover organic functional group transformations. The seventh volume is the index. The major sections are:

Volume 1. Synthesis: Carbon with No Attached Heteroatoms
Volume 2. Synthesis: Carbon with One Heteroatom Attached by a Single Bond
Volume 3. Synthesis: Carbon with One Heteroatom Attached by a Multiple Bond
Volume 4. Synthesis: Carbon with Two Heteroatoms, Each Attached by a Single Bond
Volume 5. Synthesis: Carbon with Two Attached Heteroatoms with at least One Carbon-to-Heteroatom Multiple Link
Volume 6. Synthesis: Carbon with Three or Four Attached Heteroatoms
Volume 7. Indexes

B.10.

Comprehensive Organic Transformations: A Guide to Functional Group Preparations

QD262.L355

- **functional group transformations**
- **reviews**

The reactions are general in scope and use readily available or easily prepared reagents.

EASY ACCESS

— Find the desired function group transformation in the detailed Table of Contents in the front of the book. Reactions are organized according to the functional group being prepared.
— Consult the Transformation Index in the back of the book.

INDEXING

The **Transformation Index** is a table of functional group transformations where the product molecule is referred to by a combination of family names, i.e., alkene, amide. The product (**To**) is listed first followed by the starting material (**From**) and finally the **Page** number where the transformation can be found. The names of key functional groups have been shortened, e.g., trimethylsilyl to silyl, and the halogens are not differentiated. Prefixes for all names are listed in alphabetical order and suffixes are assigned according to the principal of functional group (cation > acid > ester > acid halide > amide > nitrile > aldehyde > ketone > alcohol).

CD-ROM

Comprehensive Organic Transformations on CD-ROM: A Guide to Functional Group Preparation contains all the entries in the printed edition. It can be searched by functional group, reagent, or reaction.

ORGANIZATION

Reviews and general references are listed at the beginning of each chapter or section. The chapters contain all the reactions that give the same functional group product and are sectioned into major processes, such as elimination, reactions of a particular starting functional group, name reactions, etc.

B.11.

CRC Handbook of Chemistry and Physics:

QC61.H3

- **compound name and structure**
- **physical and chemical properties**
- **conversion and mathematical tables**

B.11.a.   OVERVIEW

The Handbook contains many tables of data useful in connection with traditional laboratory techniques (melting point, boiling point, solubility, azeotropes) as well as thermodynamic properties and mathematical information.

EASY ACCESS

— Organic (Section 3) compounds are listed alphabetically by their common name; there is a Synonym Index, Molecular Formula Index, and CAS Number Index at the end of this section.
— Consult the index (Section 1) in the back of the Handbook to locate the desired table.

INDEXING

The Index (Section 1) is located in the back of the Handbook and indexes pertinent properties, general concepts, classes of substances, and some common substances.

In **Section 3: Physical Constants of Organic Compounds**, the organic compounds are listed alphabetically by their primary name or CAS Index Name. Following this table is the *Structure Diagrams for the Table of Physical Constants* in numerical order as they occur in the table. Also included in this section is a *Synonym Index*, which includes synonyms for chemical compounds in alphabetical order with their Table identification number. The primary name for each compound is not included in this index. The *Molecular Formula Index* lists the molecular formula of chemical substances with their Table identification number and chemical name. Formulas are arranged by citing carbons in ascending order, then hydrogens, and finally the remaining elements in alphabetical order. The *CAS Number Index of Organic Compounds* at the end of this section lists the CAS Registry Numbers of all the compounds contained in the table in ascending order with their Table identification number.

CD-ROM

The *CRC Handbook of Chemistry and Physics CRCnetBASE 1999* contains all the tables and reference sections contained in the print version of the 79th edition. It can be searched by text or word with cross-table searching to afford all the information on a particular topic.

INTERNET

All the tables removed from the *CRC Handbook of Chemistry and Physics*, 71st–79th Editions can be found at:

http://www.indiana.edu/~cheminfo/crc_xtabs_71-79.html

ORGANIZATION

The Handbook is divided into 16 sections numbered with the section number in boldface followed by the page number (3-133 is Section 3, page 133). The titles of the sections and subsections are listed in B.11.b

EXAMPLE ENTRY

PHYSICAL CONSTANTS OF ORGANIC COMPOUNDS (continued)

No.	Name / Synonym	Mol. Form. / Mol. Wt.	CAS RN / mp/°C	Merck No. / bp/°C	Bell. Ref. / den/g cm^{-3}	Solubility / n_D
	Dipropyl-nitrosamine	130.19		206; 113^{40}	0.9163^{20}	1.4437^{20}
9826	1-Propanamine, *N*-propyl- Dipropylamine	$C_6H_{15}N$ 101.19	142-84-7 −63	3350 109.3	4-04-00-00469 0.7400^{20}	H_2O 3; EtOH 3; eth 5; ace 4 1.4050^{20}

Reprinted with permission from *CRC Handbook of Chemistry and Physics*. Copyright CRC Press, Boca Raton, Florida.

B.11.b.   INDEX TO THE TABLES OF THE CRC HANDBOOK

Reprinted with permission from *CRC Handbook of Chemistry and Physics*. Copyright CRC Press, Boca Raton, Florida.

Note: Major sections are in capital letters and APX indicates the appendix.

A

Abbreviations and Symbols, Scientific, Section 2
Abbreviations, Biochemical, Section 2
Absorption and Velocity of Sound in Still Air, Section 14
Absorption, Infrared, by the Earth's Atmosphere, Section 14
Absorption Properties and Neutron Scattering, Section 11
Abundance of Elements in the Earth's Crust and in the Sea, Section 14
Acceleration Due to Gravity, Section 14
Acid-Base Indicators, Section 8
Acids and Bases, Inorganic, Dissociation Constants of, Section 8
Acids, Bases, and Salts, Activity Coefficients of, Section 5
Acids, Bases, and Salts, Standard Solutions of, Section 8
Acids, Enthalpy of Dilution, Section 5
ACOUSTICS; Section 14:
Activity Coefficients of Acids, Bases, and Salts, Section 5
Activity Coefficients of Electrolytes as a Function of Concentration, Section 5
Affinities, Electron, Section 10
Air, Absorption and Velocity of Sound in, Section 14
Air, Index of Refraction of, Section 10
Air, Thermodynamic Properties of, Section 6
Air to Vacuo, Reduction of Weighings, Section 8
Air, Velocity of Sound in, Section 14
Airborne Contaminants, Threshold Limit Values for, Section 16
Allotropes of the Elements, Crystal Structures and Lattice Parameters of, Section 12
Alloys and Metals, Section 12
Alloys, Electrical Resistivity of, Section 12
Alloys, Magnetic, Properties of, Section 12
Alloys, Thermal Conductivity as a Function of Temperature, Section 12
Amino Acids, Properties of, Section 7
Amino Acids, Structures of, Section 7
ANALYTICAL CHEMISTRY; Section 8:
Analytical Reagents, Organic, for the Determination of Inorganic Substances, Section 8

Analyzing Crystals, Lattice Spacing of, Section 12
Angular Functions, Relation of, in Terms of One Another, APX A
Antiferroelectric Crystals, Properties of, Section 12
Approximate pH Values of Biological Materials and Foods, Section 7
Aqueous Electrolyte Solutions, Oxygen Solubility, Section 6
Aqueous Sodium Chloride Solutions, Volumetric Properties of, Section 6
Aqueous Solution, Conductivity Equivalent of Electrolytes in, Section 5
Aqueous Solution, Equivalent Conductivity of Electrolytes in, Section 5
Aqueous Solutions of Hydrohalogen Acids, Equivalent Conductivity of, Section 5
Aqueous Solution of Salts, Lowering of Vapor Pressure, Section 5
Aqueous Solutions, pH Scale for, Section 8
Aqueous Solutions, Viscosity of, Section 6
Astronomical Constants, Section 14
Astronomy, and Acoustics, Section 14
Atmosphere, Standard, Cloudless Sky, Total Monthly Solar Radiation in, Section 14
Atmospheric Chemistry, Kinetic and Photochemical Data for, Section 5
Atmospheric Concentration of Carbon Dioxide, 1958–1990, Section 14
Atmospheric Electricity, Section 14
Atomic Abundances, Section 1
Atomic and Molecular Polarizabilities, Section 10
Atomic Energy Levels, X-Ray, Section 10
Atomic ions, Ionization Potentials of, Section 10
Atomic Masses and Abundances, Section 1
Atomic, Molecular, and Optical Physics, Section 10
Atomic Transition Probabilities, Section 10
Atomic Weights (1993), Section 1
Atoms and Atomic Ions, Ionization Potentials of, Section 10
Attenuation, Photon, Coefficients, Section 10
Azeotropes, Section 6

B
Bases and Acids, Inorganic, Dissociation Constants of, Section 8
Bases, and Salts, Activity Coefficients of, Section 5
Bases, and Salts, Standard Solutions of, Section 8
Basic Constants, Units, and Conversion Factors, Section 1
Baths, Low Temperature, Section 15
Bead Tests, Section 8
Bessel Functions, APX A
Beta Function, APX A
Binding Energies, Electron, of the Elements, Section 10

Biochemical Symbols and Abbreviations, Section 2
Biochemistry, Section 7
Biological Buffers, Section 7
Biological Materials and Foods, Approximate pH Values of, Section 7
Black Body Radiation, Section 10
Bodies of Mass, Moment of Inertia for, APX A
Boiling, and Critical Temperatures of the Elements, Section 4
Boiling Point, Dependence on Pressure, Section 15
Boiling Point Elevation, Ebullioscopic Constants for Calculation of, Section 15
Boiling Point of Water at Various Pressures, Section 6
Boiling Points, and Melting Points of Selected Compounds, Section 6
Bond Angles and Lengths in Gas-Phase Molecules, Section 9
Bond Lengths and Angles in Gas-Phase Molecules, Section 9
Bond Lengths in Crystalline Organic Compounds, Section 9
Bond Stretching, Force Constants for, Section 9
Buffer Solutions Giving Round Values of pH at 25°C, Section 8
Buffers, Biological, Section 7

C
^{13}C Chemical Shifts of Useful NMR Solvents, Section 8
^{13}C NMR Absorptions of Major Functional Groups, Section 9
Calcite, Fluorite, and Quartz, Index of Refraction of, Section 10
Calculation of Boiling Point Elevation, Ebullioscopic Constants for, Section 15
Calculation of Freezing Point Depression, Cryoscopic Constants for, Section 15
Calibrating Conductivity Cells, Standard Solutions for, Section 5
Calibration Purposes, Index of Refraction of Liquids for, Section 10
Carbohydrates, Properties of, Section 7
Carbon Dioxide, Atmospheric Concentration of, 1958–1990, Section 14
Carbon Dioxide, Solubility in Water at Various Temperatures and Pressures, Section 6
Carcinogens, Chemical, Section 16
Carrier Gases for Gas Chromatography, Properties of, Section 8
CAS Registry Number Index of Inorganic Compounds, Section 4
CAS Registry Number Index, Organic Compounds, Section 3
Ceramics and Other Insulating Materials, Thermal Conductivity of, Section 12
Ceramics, Hardness of, Section 12
Characteristic NMR Spectral Positions for H in Organic Structures, Section 9
Characteristics of Illuminants, Section 10

Characteristics of Laser Sources, Section 10
Characteristics of Particles and Particle Dispersoids, Section 15
Characterization, Techniques for Materials, Section 12
Chemical Bonds, Strengths of, Section 9
Chemical Carcinogens, Section 16
Chemical Kinetics, Conversion Factors for, Section 1
Chemical Quantities, Symbols and Terminology, Section 2
Chemical Substances, Standard Thermodynamic Properties of, Section 5
Chemicals in Laboratories, Handling and Disposal of, Section 16
Chi-Square Distribution, Percentage Points, APX A
Chromatography, Gas, Properties of Carrier Gases for, Section 8
Chromatography, Liquid, Solvents for, Section 8
Classification of Comparative Life Hazards of Gases and Vapors, Section 16
Classification of Electromagnetic Radiation, Section 10
Clear Fused Quartz, Physical Constants of, Section 12
Cloudless Sky Standard Atmosphere, Total Monthly Solar Radiation in, Section 14
CODATA Key Values for Thermodynamics, Section 5
Code, Genetic, Section 7
Coefficient of Friction, Section 15
Coefficients, Photon Attenuation, Section 10
Coefficients, Virial, of Selected Gases, Section 6
Combustion Enthalpy of Selected Organic Compounds, Section 5
Combustion Modeling, Kinetic Data for, Section 5
Commercial Alloys and Metals, Section 12
Commercial Metals and Alloys, Section 12
Common Logarithms and their Exponential and Hyperbolic Functions, APX A
Common Optical Materials, Transmission of Light, Section 10
Common Solvents, Important Peaks in the Mass Spectra of, Section 8
Comparative Life Hazards of Gases and Vapors, Classification of, Section 16
Compressibility of Liquids, Isothermal, Section 6
Concentration, Atmospheric, of Carbon Dioxide, 1958–1990, Section 14
Concentration Function of Mean Activity Coefficients of Electrolytes, Section 5
Concentration of Solutions, Conversion Formulas for, Section 8
Conductivity Cells, Standard Solutions for Calibrating, Section 5
Conductivity Equivalent of Aqueous Solutions of Hydrohalogen Acids, Section 5

Conductivity Equivalent of Electrolytes in Aqueous Solution, Section 5

Conductivity, Ionic and Diffusion at Infinite Dilution, Section 5

Conductivity of Liquids, Thermal, Section 6

Conductivity of Gases, Thermal, Section 6

Conductivity, Thermal, of Alloys as a Function of Temperature, Section 12

Conductivity, Thermal, of Crystalline Dielectrics, Section 12

Conductivity Units, Conversion Factors for, Section 1

Configuration, Electron, of Neutral Atoms in the Ground State, Section 1

Constant Humidity Solutions, Section 15

Constants, Miscellaneous Mathematical, APX A

Contaminants, Airborne, Threshold Limit Values for, Section 16

Conversion Between Temperature Scales, Section 1

Conversion Factors, Section 1

Conversion Factors for Chemical Kinetics, Section 1

Conversion Factors for Electrical Resistivity Units, Section 1

Conversion Factors for Energy Units, Section 1

Conversion Factors for Ionizing Radiation, Section 1

Conversion Factors for Pressure Units, Section 1

Conversion Factors for Thermal Conductivity Units, Section 1

Conversion Formulas for Concentration of Solutions, Section 8

Conversion of Temperatures from the 1948 and 1968 Scales to ITS-90, Section 1

Correlation Charts, Infrared, Section 9

Cosmic Radiation, Section 11

Critical Constants, Boiling Points, and Melting Points of Selected Compounds, Section 6

Critical Temperatures of the Elements, Section 4

Cryogenic Fluids, Properties of, Section 6

Cryoscopic Constants for Calculation of Freezing Point Depression, Section 15

Crystal Lattice Energy and the Madelung Constant, Section 12

Crystal Structures and Lattice Parameters of Allotropes of the Elements, Section 12

Crystalline Dielectrics, Thermal Conductivity of, Section 12

Crystalline Organic Compounds, Bond Lengths in, Section 9

Crystals, Analyzing, Lattice Spacing of, Section 12

Crystals, Antiferroelectric Properties of, Section 12

Crystals, Cubic, Lattice Constants for, Section 12

Crystals, Elastic Constants of, Section 12

Crystals, Ferroelectric, Curie Temperature of, Section 12

Crystals, Ionic Radii, Section 12

Crystals, Symmetry of, Section 12
Cubic Crystals, Lattice Constants for, Section 12
Curie Temperature of Selected Ferroelectric Crystals, Section 12

D
Daily Dietary Allowances, Section 7
Density and Specific Volume of Mercury, Section 6
Density of D_2O, Section 6
Density of Liquid Elements, Section 4
Density of Various Solids, Section 15
Density of Water, Section 6
Density, Pressure, and Gravity as a Function of Depth within the Earth, Section 14
Dependence of Boiling Point on Pressure, Section 15
Depth within the Earth, Density, Pressure, and Gravity as a Function of, Section 14
Derivatives, APX A
Determination of Inorganic Substances, Organic Analytical Reagents for, Section 8
Deuterium Oxide, Density, Section 6
Deuterium Oxide, Fixed Point Properties of, Section 6
Deuterium Oxide, Thermal Conductivity of, Section 6
Dew Point, and Relative Humidity, Section 15
Dielectric Constant of Inorganic Solids, Section 12
Dielectric Constant of Liquids, Temperature Dependence of the, Section 6
Dielectric Constant (Permittivity) of Gases, Section 6
Dielectric Constant (Permittivity) of Liquids, Section 6
Dielectric Constants of Some Plastics and Rubbers, Section 13
Dielectric Constants of Glasses, Section 12
Dielectrics, Crystalline, Thermal Conductivity of, Section 12
Dietary Allowances, Recommended Daily, Section 7
Differential Equations, APX A
Diffusion and Ionic Conductivity at Infinite Dilution, Section 5
Diffusion Coefficients in Liquids at Infinite Dilution, Section 6
Diffusion Data for Semiconductors, Section 12
Diffusion in Gases, Section 6
Diffusivities of Metallic Solutes in Molten Metals, Section 6
Dilution of Acids, Enthalpy of, Section 5
Dimensions, and Other Parameters of the Earth, Section 14
Dipole Moments of Molecules in the Gas Phase, Section 9
Dispersoids, Particle, Characteristics of, Section 15
Disposal of Chemicals in Laboratories, Section 16

Dissociation Constants of Inorganic Acids and Bases, Section 8
Doppler Linewidths, Nomograph and Table for, Section 10
Dry Bulb Thermometer, Relative Humidity from, Section 15
Drying Agents, Efficiency of, Section 15

E

Earth, Density, Pressure, and Gravity as a Function of Depth within, Section 14
Earth, Mass, Dimensions, and Other Parameters of, Section 14
Earth's Atmosphere, Infrared Absorption by, Section 14
Earth's Crust and in the Sea, Abundance of Elements in, Section 14
Ebullioscopic Constants for Calculation of Boiling Point Elevation, Section 15
Efficacies and Other Characteristics of Illuminants, Section 10
Efficiency of Drying Agents, Section 15
Elastic Constants of Single Crystals, Section 12
Elasto-Optic, Electro-Optic, and Magneto-Optic Constants, Section 12
Electrical Resistivity of Pure Metals, Section 12
Electrical Resistivity of Selected Alloys, Section 12
Electrical Resistivity Units, Conversion Factors for, Section 1
Electricity, Atmospheric, Section 14
Electro-Optic, and Magneto-Optic Constants, Section 12
Electrochemical Series, Section 8
Electrochemistry, and Kinetics, Section 5
Electrolyte Solutions, Aqueous, Oxygen Solubility, Section 6
Electrolytes, Enthalpy of Solution, Section 5
Electrolytes in Aqueous Solution, Conductivity Equivalent of, Section 5
Electrolytes in Aqueous Solution, Equivalent Conductivity of, Section 5
Electrolytes, Mean Activity Coefficients as a Function of Concentration, Section 5
Electromagnetic Radiation, Classification of, Section 10
Electron Affinities, Section 10
Electron Binding Energies of the Elements, Section 10
Electron Configuration of Neutral Atoms in the Ground State, Section 1
Electron Emission, Secondary, Section 12
Electron Work Functions of the Elements, Section 12
Elements, Section 4
Elements, Abundance of, in the Earth's Crust and in the Sea, Section 14
ELEMENTS AND INORGANIC COMPOUNDS; Section 4:
Elements, Crystal Structures and Lattice Parameters of Allotropes of, Section 12
Elements, Electron Work Functions of, Section 12
Elements, Heat Capacity at 25 °C, Section 4

Elements, Line Spectra of, Section 10

Elements, Liquid, Density of, Section 4

Elements, Melting, Boiling, and Critical Temperatures of, Section 4

Elements, Metallic, Vapor Pressure of, Section 4

Elevation, Boiling Point, Ebullioscopic Constants for Calculation of, Section 15

Emission, Secondary Electron, Section 12

Emissivity of Total Radiation for Various Materials, Section 10

Emissivity of Tungsten, Section 10

Emissivity, Spectral, Section 10

Emissivity, Spectral, of Oxides, Section 10

Energies, Electron Binding, of the Elements, Section 10

Energies, Lattice, Section 12

Energy, Gibbs, Formation of Metal Oxides, Section 5

Energy Levels, Atomic, X-Ray, Section 10

Energy Units, Conversion Factors for, Section 1

Enthalpy of Combustion of Selected Organic Compounds, Section 5

Enthalpy of Dilution of Acids, Section 5

Enthalpy of Fusion, Section 6

Enthalpy of Solution of Electrolytes, Section 5

Enthalpy of Vaporization, Section 6

Enthalpy of Vaporization of Water, Section 6

Equations for Radiometric and Photometric Quantities, Section 10

Equivalent Conductivity of Aqueous Solutions of Hydrohalogen Acids, Section 5

Error Function, APX A

Exponential and Hyperbolic Functions and their Common Logarithms, APX A

F

F-Distribution, Percentage Points, APX A

Far-Infrared Absorption Frequency Standards, Section 10

Fats and Oils, Section 7

Fatty Acids, Properties of, Section 7

Ferroelectric Crystals, Curie Temperature of, Section 12

Fixed Point Properties of H_2O and D_2O, Section 6

Flame and Bead Tests, Section 8

Fluid Properties, Section 6

Fluids, Cryogenic, Properties of, Section 6

Fluids, Thermophysical Properties of, Section 6

Fluids, Vapor Pressure at Temperatures below 300°K, Section 6

Fluorescent Indicators, Section 8

Fluorite, and Quartz, Index of Refraction of, Section 10

Foods, Approximate pH Values of, Section 7
Force Constants for Bond Stretching, Section 9
Formation of Metal Oxides, Gibbs Energy, Section 5
Formula Index, Organic Compounds, Section 3
Formulas for Concentration of Solutions, Section 8
Fourier Series, APX A
Fourier Transforms, APX A
Freezing Point Depression, Cryoscopic Constants for Calculation of, Section 15
Freezing Points, Influence of Pressure, Section 6
Frequencies, Fundamental Vibrational, of Small Molecules, Section 9
Frequencies, Infrared Laser, Section 10
Frequency Standards, Infrared and Far-Infrared Absorption, Section 10
Friction, Coefficient of, Section 15
Functional Groups, ^{13}C NMR Absorptions of, Section 9
Fundamental Physical Constants, Section 1
Fundamental Vibrational Frequencies of Small Molecules, Section 9
Fused Quartz, Index of Refraction of, Section 10
Fused Quartz, Physical Constants of Clear, Section 12
Fusion, Enthalpy of, Section 6

G
Gamma Function, APX
Gas Chromatography, Properties of Carrier Gases for, Section 8
Gas Constant Values in Different Unit Systems, Section 1
Gas Phase Molecules, Ionization Potentials of, Section 10
Gas Phase, Dipole Moments of Molecules in, Section 9
Gas Phase Molecules, Bond Lengths and Angles, Section 9
Gases and Vapors, Classification of Comparative Life Hazards of, Section 16
Gases, Van der Waals' Constants, Section 6
Gases, Diffusion in, Section 6
Gases, Heat Capacity at 25 °C, Section 6
Gases, Mean Free Path, Section 6
Gases, Permittivity (Dielectric Constant), Section 6
Gases, Solubility in Water, Section 6
Gases, Thermal Conductivity of, Section 6
Gases, Virial Coefficients of, Section 6
Gases, Viscosity of, Section 6
Genetic Code, Section 7
Geological Time Scale, Section 14
Geophysics, Astronomy, and Acoustics, Section 14
Gibbs Energy of Formation of Metal Oxides, Section 5

Glass, Index of Refraction of, Section 10
Glass Transition Temperature for Selected Polymers, Section 13
Glasses, Dielectric Constants of, Section 12
Glasses, Thermal Conductivity of, Section 12
Gravity, Acceleration Due to, Section 14
Gravity as a Function of Depth within the Earth, Section 14
Greek, Russian, and Hebrew Alphabets, Section 2

H

Halocarbon Refrigerants, Section 6
Handling and Disposal of Chemicals in Laboratories, Section 16
Hardness of Minerals and Ceramics, Section 12
Health and Safety Information, Section 16
Heat Capacity of Liquids and Gases at 25 °C, Section 6
Heat Capacity of Selected Solids, Section 12
Heat Capacity of the Elements at 25 °C, Section 4
Heavy Water, Ionization Constant of, Section 8
Hebrew Alphabet, Section 2
High Temperature Superconductors, Section 12
Hormones, Steroid, and Other Steroidal Synthetics, Section 7
Humidity and Dew Point, Section 15
Humidity from Wet and Dry Bulb Thermometer, Section 15
Humidity Solutions, Constant, Section 15
Hydrohalogen Acids, Equivalent Conductivity of Aqueous Solutions,
    Section 5
Hyperbolic Functions and Their Common Logarithms, APX A

I

Ice, Vapor Pressure of, Section 6
Illuminants, Efficacies and Other Characteristics of, Section 10
Important Peaks in the Mass Spectra of Common Solvents, Section 8
Index of Refraction of Air, Section 10
Index of Refraction of Fused Quartz, Section 10
Index of Refraction of Glass, Section 10
Index of Refraction of Liquids for Calibration Purposes, Section 10
Index of Refraction of Rock Salt, Sylvine, Calcite, Fluorite, and Quartz,
    Section 10
Index of Refraction of Water, Section 10
Indicators, Acid-Base, Section 8
Indicators, Fluorescent, Section 8
Inertia, Moment of, for Various Bodies of Mass, APX A
Infinite Dilution, Diffusion Coefficients in Liquids at, Section 6
Infinite Dilution, Ionic Conductivity and Diffusion, Section 5

Influence of Pressure on Freezing Points, Section 6
Infrared Absorption by the Earth's Atmosphere, Section 14
Infrared and Far-Infrared Absorption Frequency Standards, Section 10
Infrared Correlation Charts, Section 9
Infrared Laser Frequencies, Section 10
Inorganic Acids and Bases, Dissociation Constants of, Section 8
Inorganic Compounds, Section 4
Inorganic Compounds, Physical Constants of, Section 4
Inorganic Compounds, Synonym Index of, Section 4
Inorganic Ions and Radicals, Nomenclature for, Section 2
Inorganic Solids, Permittivity (Dielectric Constant) of, Section 12
Inorganic Substances and Minerals, X-Ray Crystallographic Data, Section 4
Inorganic Substances, Organic Analytical Reagents for the Determination of, Section 8
Insulating Materials, Thermal Conductivity of, Section 12
Intake of Radionuclides, Section 16
Integrals, APX A
Integration, APX A
International System of Units (SI), Section 1
International Temperature Scale of 1990 (ITS-90), Section 1
Ion Exchange Resins, Section 8
Ion Product of Water Substance, Section 8
Ion Radicals, Reduction and Oxidation Potentials for, Section 8
Ionic Conductivity and Diffusion at Infinite Dilution, Section 5
Ionic Radii in Crystals, Section 12
Ionization Constant of Normal and Heavy Water, Section 8
Ionization Potentials of Atoms and Atomic Ions, Section 10
Ionization Potentials of Gas-Phase Molecules, Section 10
Ionizing Radiation, Conversion Factors for, Section 1
Ions and Radicals, Inorganic, Nomenclature for, Section 2
Irradiance, Spectral, Solar, Section 14
Isothermal Compressibility of Liquids, Section 6
Isotopes, Table of, Section 11
ITS-90, Section 1
ITS-90 Thermocouple Tables, Section 15
IUPAC Recommended Data for Vapor Pressure Calibration, Section 6

K

Kinetic and Photochemical Data for Atmospheric Chemistry, Section 5
Kinetic Data for Combustion Modeling, Section 5
Kinetics, Section 5
Kinetics, Conversion Factors for, Section 1

L

Laboratories, Handling and Disposal of Chemicals in, Section 16
Laboratory Data, Section 15
Laboratory Solvents, Properties of, Section 15
Langevin Function Values, Section 12
Laser, Infrared Frequencies, Section 10
Laser Sources, Characteristics of, Section 10
Lattice Constants for Cubic Crystals, Section 12
Lattice Energies, Section 12
Lattice Energy, Crystal, and the Madelung Constant Section 12
Lattice Parameters of Allotropes of the Elements, Section 12
Lattice Spacing of Common Analyzing Crystals, Section 12
Life Hazards of Gases and Vapors, Comparative, Classification of, Section 16
Light Transmission by Common Optical Materials, Section 10
Limits of Superheat of Pure Liquids, Section 6
Line Spectra of the Elements, Section 10
Linewidths, Doppler, Nomograph and Table for, Section 10
Liquid Baths, Low Temperature, Section 15
Liquid Chromatography, Solvents for, Section 8
Liquid Elements, Density of, Section 4
Liquid Metals, Viscosity of, Section 6
Liquids and Gases at 25 °C, Heat Capacity of, Section 6
Liquids Diffusion Coefficients at Infinite Dilution, Section 6
Liquids, Index of Refraction for Calibration Purposes, Section 10
Liquids, Isothermal Compressibility of, Section 6
Liquids, Viscosity of, Section 6
Liquids, Permittivity (Dielectric Constant), Section 6
Liquids, Surface Tension of, Section 6
Liquids, Thermal Conductivity of, Section 6
Liquids, Ultraviolet Spectra of, Section 9
Logarithms, and Their Exponential and Hyperbolic Functions, APX A
Low Temperature Liquid Baths, Section 15
Lowering of Vapor Pressure by Salts in Aqueous Solution, Section 5

M

Madelung Constant and Crystal Lattice Energy, Section 12
Magnetic Alloys, Properties of, Section 12
Magneto-Optic Constants, Section 12
Mass, Dimensions, and Other Parameters of the Earth, Section 14
Mass, Moment of Inertia for Various Bodies of, APX A
Mass Spectra of Common Solvents, Section 8
Materials Characterization Techniques, Section 12

Materials, Optical, Transmission of Light by, Section 10
Materials, Refractory, Section 12
Mathematical Constants, APX A
Mathematical Tables, Appendix A
Mean Activity Coefficients of Electrolytes as a Function of Concentration, Section 5
Mean Free Path in Gases, Section 6
Mean Temperatures in the United States, 1901–1987, Section 14
Measurements, pH, on Natural Waters, Section 8
Media, Velocity of Sound in, Section 14
Melting, Boiling, and Critical Temperatures of the Elements, Section 4
Melting Points of Selected Compounds, Section 6
Mercury, Density and Specific Volume of, Section 6
Mercury, Thermal Properties of, Section 6
Metal Oxides, Gibbs Energy of Formation, Section 5
Metallic Elements, Vapor Pressure of, Section 4
Metallic Solutes, Diffusivities in Molten Metals, Section 6
Metals and Alloys, Section 12
Metals and Semiconductors, Optical Properties of, Section 12
Metals and Semiconductors, Thermal Conductivity as a Function of Temperature, Section 12
Metals, Electrical Resistivity of, Section 12
Metals, Liquid, Viscosity of, Section 6
Metals, Molten, Diffusivities of Metallic Solutes, Section 6
Metals, Rare Earth, Physical Properties of, Section 4
Metals, Thermal and Physical Properties of, Section 12
Minerals and Ceramics, Hardness of, Section 12
Minerals, Physical and Optical Properties of, Section 4
Minerals, X-Ray Crystallographic Data, Section 4
Miscellaneous Mathematical Constants, APX A
Modeling Combustion, Kinetic Data for, Section 5
Molecular, and Optical Physics, Section 10
Molecular Formula Index, Organic Compounds, Section 3
Molecular Polarizabilities, Section 10
Molecular Structure and Spectroscopy, Section 9
Molecules, Dipole Moments in the Gas Phase, Section 9
Molecules, Fundamental Vibrational Frequencies of, Section 9
Molecules, Gas-Phase, Bond Lengths and Angles in, Section 9
Molecules, Gas-Phase, Ionization Potentials of, Section 10
Molten Metals, Diffusivities of Metallic Solutes, Section 6
Monthly Solar Radiation in a Cloudless Sky Standard Atmosphere, Section 14
Musical Scales, Section 14

N

Naming Organic Polymers, Section 13
Natural Trigonomic Functions to Four Places, APX A
Natural Width of X-Ray Lines, Section 10
Neutron Scattering and Absorption Properties, Section 11
NMR Absorptions, ^{13}C of Major Functional Groups, Section 9
NMR Solvents, ^{13}C Chemical Shifts of, Section 8
NMR Spectral Positions for H in Organic Structures, Section 9
NMR Spectroscopy, Nuclear Spins, Moments, and Other Data, Section 9
Nomenclature, Section 2
Nomenclature for Inorganic Ions and Radicals, Section 2
Nomograph and Table for Doppler Linewidths, Section 10
Nonlinear Optical Constants, Section 12
Normal Probability Function, APX A
Nuclear and Particle Physics, Section 11
Nuclear Moments, and Other Data Related to NMR Spectroscopy, Section 9
Nuclear Spins, Moments, and Other Data Related to NMR Spectroscopy, Section 9

O

Oils and Fats, Section 7
Optical Constants, Nonlinear, Section 12
Optical Materials, Transmission of Light by, Section 10
Optical Physics, Section 10
Optical Properties of Metals and Semiconductors, Section 12
Optical Properties of Minerals, Section 4
Organic Analytical Reagents for the Determination of Inorganic Substances, Section 8
Organic Compounds, Section 3
Organic Compounds, Crystalline, Bond Lengths in, Section 9
Organic Pollutants, Large Production and Properties of, Section 16
Organic Polymers, Naming, Section 13
Organic Radicals and Ring Systems, Section 2
Organic Structures, Characteristic NMR Spectral Positions for H in, Section 9
Organic Superconductors, Section 12
Orthogonal Polynomials, APX A
Oxidation and Reduction Potentials for Certain Ion Radicals, Section 8
Oxidation and Reduction Reagents, Standard Solutions of, Section 8
Oxides, Metal, Gibbs Energy of Formation, Section 5

Oxides, Spectral Emissivity of, Section 10
Oxygen Solubility in Aqueous Electrolyte Solutions, Section 6

P
Parameters of the Earth, Section 14
Particle Dispersoids, Characteristics of, Section 15
Particle Physics, Section 11
Particle Properties, Summary Tables of, Section 11
Particles and Particle Dispersoids, Characteristics of, Section 15
Percentage Points, Chi-Square Distribution, APX A
Percentage Points, F-Distribution, APX A
Percentage Points, Student's t-Distribution, APX A
Periodic Table of the Elements, Section 1
Permissible Quarterly Intake of Radionuclides, Section 16
Permittivity (Dielectric Constant) of Gases, Section 6
Permittivity (Dielectric Constant) of Inorganic Solids, Section 12
Permittivity (Dielectric Constant) of Liquids, Section 6
Permittivity (Dielectric Constant) of Liquids, Temperature Dependence
    of the, Section 6
pH, Buffer Solutions Giving Round Values, Section 8
pH Measurements on Natural Waters, Section 8
pH Scale for Aqueous Solutions, Section 8
pH Values of Biological Materials and Foods, Section 7
Photochemical and Kinetic Data for Atmospheric Chemistry, Section 5
Photometric and Radiometric Quantities, Units, Symbols, and Equations
    for, Section 10
Photon Attenuation Coefficients, Section 10
Physical and Chemical Quantities, Symbols and Terminology, Section 2
Physical and Optical Properties of Minerals, Section 4
Physical Constants, Section 1
Physical Constants of Clear Fused Quartz, Section 12
Physical Constants of Inorganic Compounds, Section 4
Physical Constants of Organic Compounds, Section 3
Physical Constants for Organic Compounds, Structure Diagrams for,
    Section 3
Physical Properties of Pure Metals, Section 12
Physical Properties of the Rare Earth Metals, Section 4
Planets, Satellites of, Section 14
Plastics and Rubbers, Dielectric Constants of, Section 13
Polarizabilities, Atomic and Molecular, Section 10
Pollutants, Organic, Large Production and Properties of, Section 16
Polymer Properties, Section 13
Polymers, Glass Transition Temperature for, Section 13

Polymers, Organic, Naming, Section 13
Polynomials, Orthogonal, APX A
Potentials, Reduction and Oxidation, for Certain Ion Radicals, Section 8
Potentials, Ionization, of Atoms and Atomic Ions, Section 10
Practical Laboratory Data, Section 15
Preparation of Reagents, Section 8
Pressure, and Gravity as a Function of Depth within the Earth, Section 14
Pressure and Temperature Solubility of Carbon Dioxide in Water, Section 6
Pressure, Dependence of Boiling Point on, Section 15
Pressure of Solids, Sublimation, Section 6
Pressure on Freezing Points, Section 6
Pressure Units, Conversion Factors for, Section 1
Pressure, Vapor, at Elevated Temperatures, Section 6
Pressure, Vapor, in the Range $-25\,°C$ to $150\,°C$, Section 6
Pressure, Vapor, of Fluids at Temperatures below $300\,°K$, Section 6
Pressure, Vapor, of Water from 0 to $370\,°C$, Section 6
Pressures, Boiling Point of Water at, Section 6
Priority Organic Pollutants, Large Production and Properties of, Section 16
Probabilities, Atomic Transition, Section 10
Probabilities, Radiative Transition, for X-Ray Lines, Section 10
Probability Function, APX A
Production and Priority Organic Pollutants, Properties of, Section 16
Properties of Antiferroelectric Crystals, Section 12
Properties of Carbohydrates, Section 7
Properties of Common Amino Acids, Section 7
Properties of Common Laboratory Solvents, Section 15
Properties of Cryogenic Fluids, Section 6
Properties of Large Production and Priority Organic Pollutants, Section 16
Properties of Magnetic Alloys, Section 12
Properties of Purine and Pyrimidine Bases, Section 7
Properties of Selected Fatty Acids, Section 7
Properties of Semiconductors, Section 12
Properties of Solids, Section 12
Properties of Superconductors, Section 12
Properties of the Elements and Inorganic Compounds, Section 4
Properties of the Solar System, Section 14
Properties of Tungsten, Section 10
Properties of Water in the Range $0-100\,°C$, Section 6
Pure Liquids, Limits of Superheat, Section 6

Purine and Pyrimidine Bases, Properties of, Section 7
Pyrimidine Bases, Properties of, Section 7

Q
Quarterly Intake of Radionuclides, Section 16
Quartz, Clear Fused, Physical Constants of, Section 12
Quartz, Fused, Index of Refraction of, Section 10
Quartz, Index of Refraction of, Section 10

R
Radiation, Black Body, Section 10
Radiation, Cosmic, Section 11
Radiation, Electromagnetic, Classification of, Section 10
Radiation, Emissivity of, Total, for Various Materials, Section 10
Radiation, Solar, in a Cloudless Sky Standard Atmosphere, Section 14
Radiative Transition Probabilities for X-Ray Lines, Section 10
Radicals and Ring Systems, Organic, Section 2
Radicals, Inorganic, Nomenclature for, Section 2
Radicals, Ion, Reduction and Oxidation Potentials for, Section 8
Radii, Ionic, in Crystals, Section 12
Radiometric and Photometric Quantities, Units, Symbols, and Equations
    for, Section 10
Radionuclides, Permissible Quarterly Intake of, Section 16
Rare Earth Metals, Physical Properties of, Section 4
Reagents, Organic Analytical, for the Determination of Inorganic
    Substances, Section 8
Reagents, Preparation of, Section 8
Recommended Daily Dietary Allowances, Section 7
Reduction and Oxidation Potentials for Certain Ion Radicals, Section 8
Reduction and Oxidation Reagents, Standard Solutions of, Section 8
Reduction of Weighings in Air to Vacuo, Section 8
Refractory Materials, Section 12
Refrigerants, Halocarbon, Section 6
Registry Number Index of Inorganic Compounds, Section 4
Registry Number Index, Organic Compounds, Section 3
Relation of Angular Functions in Terms of One Another, APX A
Relative Dew Point, Section 15
Relative Humidity and Dew Point, Section 15
Relative Humidity from Wet and Dry Bulb Thermometer, Section 15
Resins, Ion Exchange, Section 8
Resistivity, Electrical, of Pure Metals, Section 12
Resistivity, Electrical, of Selected Alloys, Section 12
Resistivity Units, Conversion Factors for, Section 1

Ring Systems, Organic, Section 2
Rock Salt, Sylvine, Calcite, Fluorite, and Quartz, Index of Refraction
    of, Section 10
Rubbers, Dielectric Constants of, Section 13
Russian and Hebrew Alphabets, Section 2

S
Safety Information, Section 16
Salts, Activity Coefficients of, Section 5
Salts in Aqueous Solution, Lowering of Vapor Pressure by, Section 5
Salts, Standard Solutions of, Section 8
Satellites of the Planets, Section 14
Scales, Musical, Section 14
Scientific Abbreviations and Symbols, Section 2
Sea and Earth's Crust, Abundance of Elements in, Section 14
Secondary Electron Emission, Section 12
Semiconductors, Diffusion Data for, Section 12
Semiconductors, Optical Properties of, Section 12
Semiconductors, Properties of, Section 12
Semiconductors, Thermal Conductivity as a Function of Temperature,
    Section 12
Series Expansion, APX A
SI Units, Section 1
Single Crystals, Elastic Constants of, Section 12
Sky, Cloudless, Standard Atmosphere, Total Monthly Solar Radiation
    in, Section 14
Solar Radiation in a Cloudless Sky Standard Atmosphere, Section 14
Solar Spectral Irradiance, Section 14
Solar System, Properties of, Section 14
Solids, Section 12
Solids, Density of, Section 15
Solids, Heat Capacity of, Section 12
Solids, Inorganic, Permittivity (Dielectric Constant) of, Section 12
Solids, Sublimation Pressure of, Section 6
Solubility Chart, Section 8
Solubility of Carbon Dioxide in Water at Various Temperatures and
    Pressures, Section 6
Solubility of Selected Gases in Water, Section 6
Solubility Product Constants, Section 8
Solutes, Metallic, Diffusivities in Molten Metals, Section 6
Solution Enthalpy of Electrolytes, Section 5
Solutions, Aqueous, Viscosity of, Section 6
Solutions, Constant Humidity, Section 15

Solutions, Conversion Formulas for Concentration of, Section 8
Solutions of Acids, Bases, and Salts, Section 8
Solutions of Oxidation and Reduction Reagents, Section 8
Solvents for Liquid Chromatography, Section 8
Solvents for Ultraviolet Spectrophotometry, Section 8
Solvents, Important Peaks in the Mass Spectra of, Section 8
Solvents, Properties of, Section 15
Sound in Still Air, Absorption and Velocity of, Section 14
Sound, Velocity of in Dry Air, Section 14
Sound, Velocity of in Various Media, Section 14
Sources, Laser, Characteristics of, Section 10
Specific Volume of Mercury, Section 6
Spectra, Line, of the Elements, Section 10
Spectra, Ultraviolet, of Common Liquids, Section 9
Spectral Emissivity, Section 10
Spectral Emissivity of Oxides, Section 10
Spectral Irradiance, Solar, Section 14
Spectrophotometry, Ultraviolet, Solvents for, Section 8
Spectroscopy, Section 9
Standard Atmosphere, Cloudless Sky, Total Monthly Solar Radiation
    in, Section 14
Standard Atomic Weights (1993), Section 1
Standard Density of Water, Section 6
Standard ITS-90 Thermocouple Tables, Section 15
Standard Solutions for Calibrating Conductivity Cells, Section 5
Standard Solutions of Acids, Bases, and Salts, Section 8
Standard Solutions of Oxidation and Reduction Reagents, Section 8
Standard Thermodynamic Properties of Chemical Substances, Section 5
Steam Tables, Section 6
Steroid Hormones and Other Steroidal Synthetics, Section 7
Steroidal Synthetics and Steroid Hormones, Section 7
Strengths of Chemical Bonds, Section 9
Stretching, Bond, Force Constants for, Section 9
Structure Diagrams for Table of Physical Constants, Section 3
Structures of Common Amino Acids, Section 7
Student's t-Distribution, Percentage Points, APX A
Sublimation Pressure of Solids, Section 6
Summary Tables of Particle Properties, Section 11
Superconductors, High Temperature, Section 12
Superconductors, Organic, Section 12
Superconductors, Properties of, Section 12
Superheat, Limits of Pure Liquids, Section 6
Surface Tension of Common Liquids, Section 6

Sylvine, Calcite, Fluorite, and Quartz, Index of Refraction of, Section 10·

Symbols and Abbreviations, Biochemical, Section 2

Symbols, and Equations for Radiometric and Photometric Quantities, Section 10

Symbols and Terminology for Physical and Chemical Quantities, Section 2

Symbols, Scientific, Section 2

Symbols, Terminology, and Nomenclature, Section 2

Symmetry of Crystals, Section 12

Synonym Index of Inorganic Compounds, Section 4

Synonym Index for Organic Compounds, Section 3

T

t-Distribution, Percentage Points, APX

Table of the Isotopes, Section 11

Table for Doppler Linewidths, Section 10

Techniques for Materials Characterization, Section 12

Temperature and Pressure Solubility of Carbon Dioxide in Water, Section 6

Temperature, Curie, of Selected Ferroelectric Crystals, Section 12

Temperature Dependence of the Permittivity (Dielectric Constant) of Liquids, Section 6

Temperature, Glass Transition for Selected Polymers, Section 13

Temperature, High, Superconductors, Section 12

Temperature Scale of 1990 (ITS-90), Section 1

Temperature Scales, Conversion Between, Section 1

Temperature, Thermal Conductivity of Alloys as a Function of, Section 12

Temperature, Thermodynamic Properties as a Function of, Section 5

Temperatures from the 1948 and 1968 Scales to ITS-90, Conversion of, Section 1

Temperatures, Mean, in the United States, 1901–1987, Section 14

Temperatures, Vapor Pressure at, Section 6

Terminology, and Nomenclature, Section 2

Terminology for Physical and Chemical Quantities, Section 2

The Madelung Constant and Crystal Lattice Energy, Section 12

Thermal and Physical Properties of Pure Metals, Section 12

Thermal Conductivity of Alloys as a Function of Temperature, Section 12

Thermal Conductivity of Ceramics and Other Insulating Materials, Section 12

Thermal Conductivity of Crystalline Dielectrics, Section 12

Thermal Conductivity of Gases, Section 6
Thermal Conductivity of Glasses, Section 12
Thermal Conductivity of Liquids, Section 6
Thermal Conductivity of Metals and Semiconductors as a Function of Temperature, Section 12
Thermal Conductivity of Saturated H_2O and D_2O, Section 6
Thermal Conductivity Units, Conversion Factors for, Section 1
Thermal Properties of Mercury, Section 6
Thermochemistry, Electrochemistry, and Kinetics, Section 5
Thermocouple Tables, Standard ITS-90, Section 15
Thermodynamic Properties as a Function of Temperature, Section 5
Thermodynamic Properties of Air, Section 6
Thermodynamic Properties of Chemical Substances, Section 5
Thermodynamics, CODATA Key Values for, Section 5
Thermometer, Relative Humidity from, Section 15
Thermophysical Properties of Fluids, Section 6
Threshold Limit Values for Airborne Contaminants, Section 16
Time Scale, Geological, Section 14
TLV for Airborne Contaminants, Section 16
Total Monthly Solar Radiation in a Cloudless Sky Standard Atmosphere, Section 14
Transition Probabilities, Atomic, Section 10
Transition Probabilities for X-Ray Lines, Section 10
Transmission of Light by Common Optical Materials, Section 10
Trigonomic Functions to Four Places, APX A
Tungsten, Emissivity of, Section 10
Tungsten, Properties of, Section 10

U
Ultraviolet Spectra of Common Liquids, Section 9
Ultraviolet Spectrophotometry, Solvents for, Section 8
United States, Mean Temperatures in, 1901–1987, Section 14
Units, and Conversion Factors, Section 1
Units, Symbols, and Equations for Radiometric and Photometric Quantities, Section 10

V
Vacuo to Air, Reduction of Weighings, Section 8
Values for the Langevin Function, Section 12
Values of the Gas Constant in Different Unit Systems, Section 1
Van der Waals' Constants for Gases, Section 6
Vapor Pressure at Elevated Temperatures, Section 6
Vapor Pressure Calibration, IUPAC Recommended Data for, Section 6

Vapor Pressure Lowering by Salts in Aqueous Solution, Section 5
Vapor Pressure in the Range −25 °C to 150 °C, Section 6
Vapor Pressure of Fluids at Temperatures below 300 °K, Section 6
Vapor Pressure of Ice, Section 6
Vapor Pressure of the Metallic Elements, Section 4
Vapor Pressure of Water from 0 to 370 °C, Section 6
Vaporization, Enthalpy of, Section 6
Vaporization of Water, Enthalpy of, Section 6
Vapors, Classification of Comparative Life Hazards of, Section 16
Vector Analysis, APX A
Velocity of Sound in Dry Air, Section 14
Velocity of Sound in Still Air, Section 14
Velocity of Sound in Various Media, Section 14
Vibrational Frequencies of Small Molecules, Section 9
Virial Coefficients of Selected Gases, Section 6
Viscosity of Aqueous Solutions, Section 6
Viscosity of Gases, Section 6
Viscosity of Liquid Metals, Section 6
Volumetric Properties of Aqueous Sodium Chloride Solutions, Section

W

Water, Boiling Point at Various Pressures, Section 6
Water, Enthalpy of Vaporization, Section 6
Water, Fixed Point Properties of, Section 6
Water, Index of Refraction of, Section 10
Water, Ionization Constant of, Section 8
Water, Properties in the Range 0−100 °C, Section 6
Water Solubility of Carbon Dioxide at Various Temperatures and Pressures, Section 6
Water Solubility of Selected Gases, Section 6
Water, Standard Density of, Section 6
Water Substance, Ion Product of, Section 8
Water, Thermal Conductivity of, Section 6
Water, Vapor Pressure from 0 to 370 °C, Section 6
Waters, Practical pH Measurements on, Section 8
Wavelengths, X-Ray, Section 10
Weighings in Air to Vacuo, Reduction of, Section 8
Wet and Dry Bulb Thermometer, Relative Humidity from, Section 15
Width of X-Ray Lines, Section 10
Wire Tables, Section 15
Work Functions, Electron, of the Elements, Section 12

X

X-Ray Atomic Energy Levels, Section 10

X-Ray Crystallographic Data on Inorganic Substances and Minerals, Section 4

X-Ray Lines, Natural Width of, Section 10

X-Ray Lines, Radiative Transition Probabilities for, Section 10

X-Ray Wavelengths, Section 10

B.12.

CRC Handbook of Data on Organic Compounds:

QD257.7.C73

- **compound name and structure**
- **physical and chemical properties**
- **spectroscopic information**

The Handbook contains many tables of data useful in the identification of unknown organic compounds. HODOC is the abbreviation for "Handbook of Data on Organic Compounds."

EASY ACCESS

— The compounds are listed alphabetically by their CAS Chemical Index name.
— The Molecular Formula Index gives the HODOC number of all compounds with a particular molecular composition.
— The Synonym Index alphabetically lists all compound synonyms with their corresponding HODOC number.

INDEXING

The **Molecular Formula Index** lists the molecular formula of chemical substances with the HODOC number of all compounds with that molecular composition. Formulas are arranged by citing carbons in ascending order, then hydrogens, and finally the remaining elements in alphabetical order.

The **Molecular Weight Index** lists the molecular weight of chemical substances in bold type to two significant digits with the HODOC number of all the compounds with that molecular weight.

The **CAS Number Index** correlates the CAS Registry number with the HODOC number for all the compounds in the Handbook.

The **Melting Point Index** lists the melting point of chemical substances in bold type with the HODOC number of all the compounds with that melting point.

The **Synonym Index** includes synonyms to chemical compounds in alphabetical order with their HODOC identification number. The primary name for each compound is not included in this index.

The **Boiling Point Index** lists the boiling point of chemical substances in bold type with the HODOC number of all the compounds with that boiling point.

The **Infrared Numerical Index of Source Curves** lists IR curves from the various reference collections that were used to create the database with the HODOC number of the corresponding compound. The abbreviations are defined in the "Explanations to Spectral Data Indexes" section of those volumes containing spectral indexes.

The **Infrared Spectral Data Index** is a listing of IR peaks from the "fingerprint" region of the IR ($1300-250$ cm^{-1}) in order of the lowest wavenumber. The HODOC number of the associated compound is listed first, followed by the IR peaks in numerical order. If the lowest wavenumber peaks are identical, the order is determined by the next highest peak observed in the spectra.

The **Infrared Special Spectral Data Indexes** lists IR peaks from the "fingerprint" region of the IR ($1300-250$ cm^{-1}), in order of the lowest wavenumber for hydrocarbons, CHO, CHN, CHNO, halogens, B, P, Si, and S compounds. The HODOC number of the associated compound is listed first, followed by the IR peaks in numerical order. If the lowest wavenumber peaks are identical, the order is determined by the next highest peak observed in the spectra.

The **Raman Numerical Index of Source Curves** lists the Raman curves from the various reference collections that were used to create the database with the HODOC number of the corresponding compound. The abbreviations are defined in the "Explanations to Spectral Data Indexes" section of those volumes containing spectral indexes.

The **Raman Spectral Data Index** is a listing of Raman bands from the "characteristic" region of $50-1300$ cm^{-1} in order of the lowest wavenumber. The HODOC number of the associated compound is listed first, followed by the Raman bands in numerical order. If the lowest wavenumber bands are identical, the order is determined by the next highest band observed in the spectra.

The **Ultraviolet Numerical Index of Source Curves** lists the UV curves from the various reference collections that were used to create the database with the HODOC number of the corresponding compound. The abbreviations are defined in the "Explanations to Spectral Data Indexes" section of those volumes containing spectral indexes. For journal articles, the journal abbreviation is followed by the volume number, page number, and year separated by hyphens.

The **Ultraviolet Spectral Data Index** is composed of three columns. The first column is the compound's HODOC number, the middle column has the wavelength of the strongest band with the extinction coefficient in parenthesis, and the last column lists the other strong bands in the spectra. Compounds are listed in decreasing order of the strongest band (middle column), with decreasing wavelength for additional bands (third column) as a subsequent criteria.

The ^{1}H **NMR Numerical Index of Source Curves** lists the proton NMR curves from the various reference collections that were used to create the database with the HODOC number of the corresponding compound. The abbreviations are defined in the "Explanations to Spectral Data Indexes" section of those volumes containing spectral indexes.

The ^{1}H **NMR Spectral Data Index** contains two different indexes: "High Field to Low" and "Low Field to High." Both indexes list the compound's HODOC number in the first column and the NMR chemical shifts in the second. The "High Field to Low" section lists the proton NMR data in increasing order starting from 0.0 δ, and the "Low Field to High" section lists the data in decreasing order from 15.5 δ.

The ^{1}H **NMR Special Spectral Data Indices** are a listing of NMR peaks for hydrocarbons, CHO, CHN, CHNO, halogens, B, P, Si, and S compounds. The data are listed the proton NMR data in decreasing order from 15.5 δ. The HODOC number of the associated compound is listed first, followed by the NMR peaks in numerical order. If the lowest wavenumber peaks are identical, the order is determined by the next low field peak observed in the spectra.

The ^{13}C **NMR Numerical Index of Source Curves** lists the carbon NMR curves from the various reference collections that were used to create the database with the HODOC number of the corresponding compound. The abbreviations are defined in the "Explanations to Spectral Data Indexes" section of those volumes containing spectral indexes.

The ^{13}C **NMR Spectral Data Index** contains two different indices: "High Field to Low" and "Low Field to High." Both indexes list the compound's HODOC number in the first column and the NMR chemical shifts in the second. The "High Field to Low" section lists the carbon NMR data in increasing order starting from 0.0 δ, and the "Low Field to High" section lists the data in decreasing order from 200.0 δ.

The **Mass Spectra Numerical Index of Source Curves** lists the mass spectrum data from the various reference collections that were used to create the database with the HODOC number of the corresponding compound. The abbreviations are defined in the "Explanations to Spectral Data Indexes" section of those volumes containing spectral indexes.

The **Mass Spectra Spectral Data Index** lists the HODOC number of the compound followed by the mass number of the most intense *m/e* peak value in the spectrum. The masses are listed in decreasing order, and the second and third most intense peaks are also given to aid in identification.

The **Mass Spectrometry: Special Molecular Weight Indexes** list *m/e* values for Br, Cl, N, and S compounds and consist of three columns. The first column is the compound's HODOC number, the middle column has the molecular weight of the compound to two significant digits, and the last column lists the *m/e* peak values beginning with the most intense in the spectra. Compounds are listed in decreasing order of molecular weight (middle column), with decreasing *m/e* peak values (third column) as a subsequent criteria.

CD-ROM

The *Properties of Organic Compounds on CD-ROM* contains 27,500 of the most commonly used organic compounds. It can be searched by molecular formula, physical or spectral property, CAS Index Name or Registry Number.

COMPUTER SEARCH

The **CRC Handbook of Data on Organic Compounds** is available through STN International (FILE HODOC). The Basic Index (BI) contains single words from the notes, solvent data, and crystal description as well as CAS Registry Numbers, molecular formulas, and chemical names. Example search fields and display codes are given below.

Search Field Name	Search Code	Search Example	Display Code
Boiling point	/BP	S BP >= 100	BP
CAS registry number	/RN	S 67-64-1/RN	RN
Chemical name	/CN	S DURACEF/CN	CN
HODOC number	/HN	S 25582/HN	HN
Melting point	/MP	S 200-300/MP	MP
Molecular formula	/MF	S CL6SI2/MF	MF

The default display format, DISPLAY L(*ist number*), in HODOC provides the CAS registry no. (RN), HODOC number (HN), molecular formula (MF), linear structural formula (LSF), CA index name (CN), and synonyms (CN). The DISPLAY ALL command also gives the references (RE), other sources i.e. Beilstein number (OS), molecular weight (MW), crystal property (CPD), solubility (SLB), refractive index (RI), boiling point (BP), melting point (MP), density (DEN), optical rotation (OPR), IR spectrum (IRS), Raman spectrum (RAS), UV and visible spectrum (UVS), NMR spectrum (NMRS), and mass spectrum (MS). This format includes all data for a given substance and pertinent references. Any of these individual data fields can be accessed with the DISPLAY command.

(Copyright 1998 by the American Chemical Society and reprinted with permission.)

INTERNET

STN HODOC Database Summary Sheets can be accessed at:

> http://info.cas.org/ONLINE/DBSS/hodocss.html

ORGANIZATION

The **CRC Handbook of Data on Organic Compounds** is an alphabetical listing of compounds by CAS Chemical Index Name. Each entry is assigned a HODOC number corresponding numerically to its position in the Handbook.

B.13.

Dictionary of Organometallic Compounds:

QD411.D53

- compound name and structure
- preparation and occurrence
- physical properties and derivatives

The following types of compounds are included: compounds representative of important structural types, catalysis, synthetic reagents, chemicals of commerce, and other compounds with interesting structural, chemical, or bonding properties. The literature citations to the leading references on a compound and limited data on derivatives are included.

EASY ACCESS

— Use the Molecular Formula Index in Volume 5 of the Main Work to find Name and Entry Number for a compound.
— If the compound cannot be found in the Index to the Main Work, consult the Molecular Formula Index in the latest Supplement.

INDEXING

Volume 5 of the Main Work and each Supplement contains cumulative Name, Molecular Formula, and Chemical Abstracts Registry Number indices. The cumulative **Name Index** lists the chemical substances with their Entry Number consisting of a metal element symbol followed by a five-digit number. The numbers are listed chronologically within each metal section.

The **Molecular Formula Index** lists the molecular formula of the chemical substances with their chemical name and Entry Number. Formulas are arranged by citing carbons in ascending order, then hydrogens, and finally the remaining elements in alphabetical order. If carbon is not present, the elements, including H, are arranged in alphabetical order. Where a molecular formula applies to a derivative, the DOC Number is prefixed by the word *in*. The symbol â preceding an Entry

Name indicates that the entry referred to contains information on toxic or hazardous properties.

The **Chemical Abstracts Service Registry Number Index** lists all CAS Registry Numbers recorded in ascending numerical order with the corresponding chemical name and Entry Number. Where a molecular formula applies to a derivative, the DOC Number is prefixed by the word *in*. The symbol â preceding an Entry Name indicates that the entry referred to contains information on toxic or hazardous properties.

The first few pages of the **Index of Synthetic Reagents** lists synthetic organic transformations that can be conducted with organometallic compounds as either catalytic or stoichiometric reagents. The transformations are listed with the page in the index where they appear. The remainder of the index links the synthetic transformation with an organometallic compound and its DOC Number.

CD-ROM

The *Dictionary of Inorganic and Organometallic Compounds on CD-ROM* contains all the entries from the **Dictionary of Organometallic Compounds** and the **Dictionary of Inorganic Compounds**. It can be searched by text, structure, or substructure.

COMPUTER SEARCH:

The **Dictionary of Organometallic Compounds** is included in the Chapman and Hall Chemical Database (CHCD) available through Dialog Information Retrieval Service (Database [303]). The Basic Index (BI) contains the Main CHCD Substance Name (CN), derivative substance information (DERIV), data present (DP), main CHCD substance information (MAIN), element count (EC), variant name (NA, DE) synonyms (SY), compound type, hazard information, general information, source of substance, use/importance, miscellaneous, physical state (TX), compound variant information (VAR). Example search fields and display codes are given below.

Search Field Name	Search Code	Search Example
Chemical name	CN=	S CN=DURACEF
CAS Registry No.	RN=	S RN=4834-58-6
Molecular formula	MF=	S MF=C3H6N2O2

The display formats are:

Format 1   DIALOG Accession Number
Format 2   Individual Compound (Variant, Derivative, or Main Compound)
Format 3   Individual Compound (Variant, Derivative, or Main Compound)
Format 4   Individual Compound (Variant, Derivative, or Main Compound)
Format 5   Full Record
Format 6   Compound Name, CAS Registry Number, Data Present
Format 7   Individual Compound (Variant, Derivative, or Main Compound)
Format 8   Compound Name, Synonyms, CAS Registry Number, Data Present, Subfile
Format 9   Full Record
Format K   KWIC (Key Word In Context) displays a window of text; may be used by itself or with other formats (HILIGHT is also available)

Copyright The Dialog Corporation 1998.

INTERNET

Dialog Information Retrieval Service Blue Sheets can be accessed at:

http://library.dialog.com/bluesheets/

ORGANIZATION

The Dictionary is divided into Sections for each of the elements except for the halogens, noble gases and H, C, N, O, P, S, Se, Te and unstable radioactive elements. Each entry in the **Dictionary of Organometallic Compounds** is given an Entry Number consisting of a metal element symbol followed by a five-digit number. In the Main Work volumes, the first digit is 0; the first Supplement, 1; the second Supplement, 2, etc.

The first page of each element section contains information on the pure element, such as analysis, atomic number, atomic weight, availability, color, coordination number, electronic configuration, foreign language names, handling, isotopic abundance, nomenclature, references, spectroscopy, toxicity, and valency. Next, the element section contains a graphical structure index that reproduces all structure diagrams for that element section and in entry-number order. Within each element section, the molecular formulas are arranged by citing carbons in ascending order, then hydrogens, and finally the remaining elements in alphabetical order. If carbon is not present, the elements, including H, are arranged in alphabetical order. The entries for compounds that contain more than one type of metal atom are printed in all relevant element sections.

EXAMPLE ENTRY

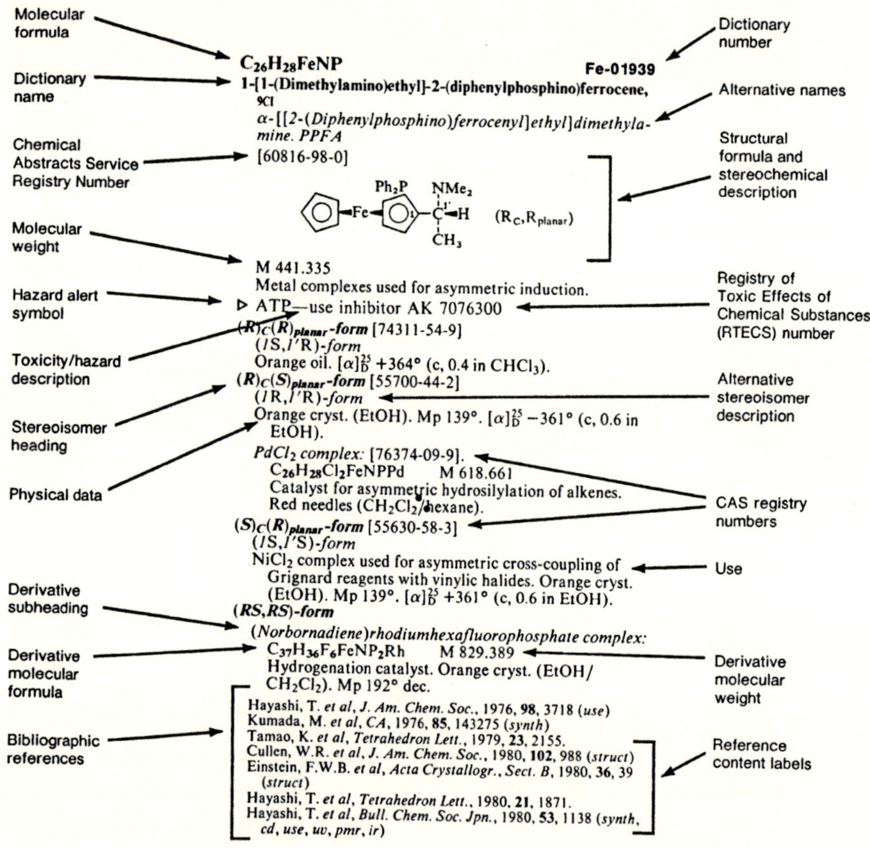

Reprinted from the *Dictionary of Organometallic Compounds* (1984), J. Buckingham (ed.) with kind permission from Kluwer Academic Publishers.

B.14.

Dictionary of Organic Compounds (Heilbron):

QD246.D5

- compound name and structure
- preparation and occurrence
- physical properties and derivatives

The following types of compounds are included: natural products, fundamental organic compounds, laboratory reagents and solvents, chemicals of commerce, and other compounds with interesting structural, chemical, or biological properties. The literature citations to the leading references on a compound and limited data on derivatives are included.

EASY ACCESS

— To find a particular compound, use the Molecular Formula Index to the Main Work to find the DOC Name and Number.
— If the compound cannot be found in the Index to the Main Work, consult the latest Supplement Cumulative Index.

INDEXING

The cumulative **Molecular Formula Index** lists the molecular formula of chemical substances with their DOC name and number. Formulas are arranged by citing carbons in ascending order, then hydrogens, and finally the remaining elements in alphabetical order. If carbon is not present, the elements are arranged in alphabetical order. Where a molecular formula applies to a derivative, the DOC Number is prefixed by the word *in*. The symbol â preceding an Entry Name indicates that the entry referred to contains information on toxic or hazardous properties.

The **Chemical Abstracts Service Registry Numbers Index** lists in ascending numerical order all CAS Registry Numbers with corresponding DOC name and number. A DOC Number that is preceded by the word *in* means that the CAS Registry Number refers to the specified stereoisomer or derivative that is to be found embedded within the particular entry. The symbol â preceding an Entry Name indicates that the entry referred to contains information on toxic or hazardous properties.

The **Heteroatom Index** lists each heteroatom (not C, H, O, N, or halogens) alphabetically by conventional element abbreviation (Ag for silver is the first heteroatom listed). Within each heteroatom section, the molecular formula of a compound are arranged by citing carbons in ascending order, then hydrogens, and finally the remaining elements in alphabetical order. If carbon is not present, the elements are arranged in alphabetical order.

Where a molecular formula applies to a derivative, the DOC Number is prefixed by the word *in*. The symbol â preceding an Entry Name indicates that the entry referred to contains information on toxic or hazardous properties.

CD-ROM

The *Dictionary of Organic Compounds on CD-ROM* contains all the entries from the **Dictionary of Organic Compounds**. It can be searched by text, structure, or substructure.

COMPUTER SEARCH

The **Dictionary of Organic Compounds** is included in the Chapman and Hall Chemical Database (CHCD) available through Dialog Information Retrieval Service (Database [303]). The Basic Index (BI) contains the Main CHCD Substance Name (CN), derivative substance information (DERIV), data present (DP), main CHCD substance information (MAIN), element count (EC), variant name (NA, DE) synonyms (SY), compound type, hazard information, general information, source of substance, use/importance, miscellaneous, physical state (TX), compound variant information (VAR). Example search fields and display codes are given below.

Search Field Name	Search Code	Search Example
Chemical name	CN=	S CN=DURACEF
CAS Registry No.	RN=	S RN=4834-58-6
Molecular formula	MF=	S MF=C3H6N2O2

The display formats are:

Format 1   DIALOG Accession Number
Format 2   Individual Compound (Variant, Derivative, or Main Compound)
Format 3   Individual Compound (Variant, Derivative, or Main Compound)
Format 4   Individual Compound (Variant, Derivative, or Main Compound)
Format 5   Full Record
Format 6   Compound Name, CAS Registry Number, Data Present
Format 7   Individual Compound (Variant, Derivative, or Main Compound)
Format 8   Compound Name, Synonyms, CAS Registry Number, Data Present, Subfile
Format 9   Full Record
Format K   KWIC (Key Word In Context) displays a window of text; may be used by itself or with other formats (HILIGHT is also available)

Copyright The Dialog Corporation 1998

INTERNET

Dialog Information Retrieval Service Blue Sheets can be accessed at:

http://library.dialog.com/bluesheets/

ORGANIZATION

Each entry in the Dictionary of Organic Compounds (DOC) is alphabetical by DOC Name and given a DOC Number consisting of a letter followed by a five-digit number. In the Main Work volumes, the first digit is 0; the first Supplement, 1; the second Supplement, 2; etc.

LITERATURE DESCRIPTION

P. H. Rhodes *The Organic Chemist's Desk Reference*. New York: Chapman & Hall Publishers, 1995.

EXAMPLE ENTRY

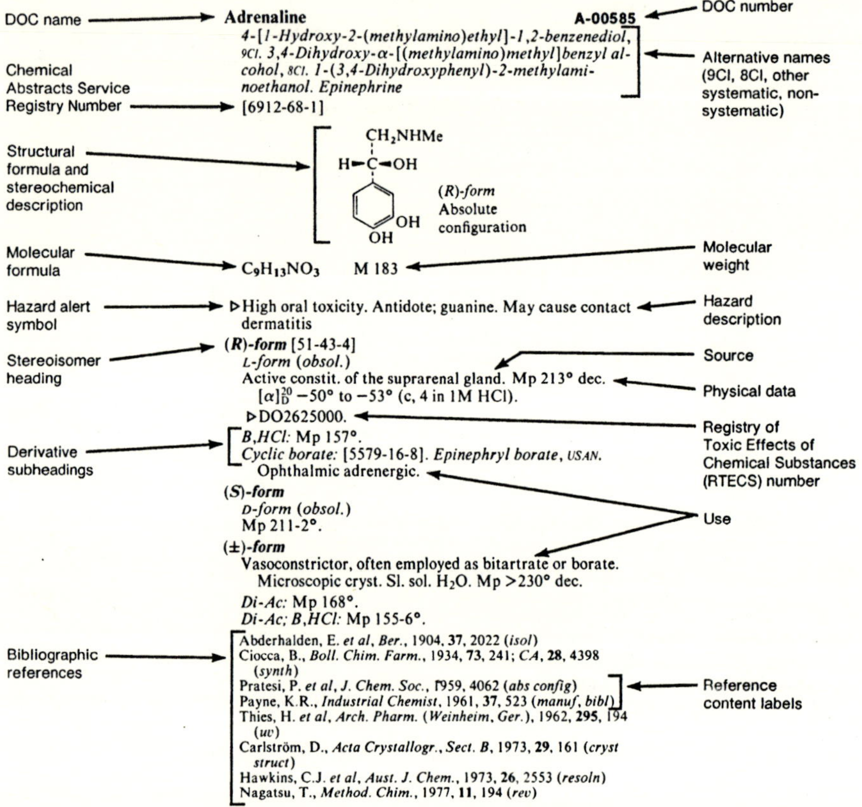

Reprinted from the *Dictionary of Organic Compounds* (1982), J. Buckingham (ed.) with kind permission from Kluwer Academic Publishers.

B.15.

Dissertations

BOOKS

Dissertation Abstracts. Ann Arbor, Mich.: University Microfilms (Z5055.U5 A53).

DATABASES

Dissertation Abstracts database is available through Dialog Information Retrieval Service (Database [35]) or through STN International (FILE DISSABS). For the STN database, the Basic Index (BI) contains single words from the title (TI) and the abstract. Example search fields and display codes are given below.

Search Field Name	Search Code	Search Example	Display Code
Accession Number	/AN	S 102:123456	AN
Author/Advisor	/AU	S GILPIN, R/AU	AU
Institution	/CS	S DUKE/CS	CS
Publication Date	/PD	S 19920000/PD	SO
Publication Year	/PY	S 1950/PY	PY
Title	/TI	S ISOMER/TI	TI

The default display format, DISPLAY L(*ist number*), in DISSABS provides information on the accession number (AN), title (TI), author/advisor (AU), institution (CS), file segment (FS) and language (LA). The DISPLAY ALL command also gives the abstract (AB) and classification code or descriptor (CC). Any of these individual data fields can be accessed with the DISPLAY command.

INTERNET RESOURCES

UMI Company, the producer of University Microfilms, Inc., has Proquest Digital Dissertations and Dissertation Express available through their web site at: http://www.umi.com. The most recent three months of UMI's Dissertation Abstracts database can be accessed online. The entire database is available by subscription.

B.16.

Encyclopedia of Reagents for Organic Synthesis

QD77.E53

- **preparation and purification of compounds**
- **reactions of organic reagents**
- **reviews**
- **physical data**

B.16.a.   OVERVIEW

A description of the utility of all reagents used in organic chemistry. The Encyclopedia contains physical data, solubility, purification, preparative methods for all reagents, or combination of reagents, which effect the transformation of a substrate into a product.

EASY ACCESS

— The Subject Index in Volume 8 provides the volume in bold print and page number for each reagent covered in the encyclopedia.

INDEXING

Volume 8 contains four separate indexes to locate information about specific organic reagents, named reagents or reactions, types of reactions, and various products or substrates.

The **Reagent Formula Index** lists the molecular formula of all reagents listed in the Encyclopedia with their names and entry volume (bold print) and page. Formulas are arranged in alphabetical order if carbon is not present. Otherwise the carbons are cited in ascending order, then hydrogens, and finally the remaining elements in alphabetical order. Any complexes or mixed reagents of a compound are listed under the molecular formula of the parent reagent.

In the **Reagent Function Index**, the first two pages list the named reactions and general types of transformations or catalysts covered by the index with the page where the category can be found. The remainder of the Function Index links the specific reagent or catalyst in each general category with the volume (bold print) and page where the entry can be found in the encyclopedia.

In the **Reagent Structural Class Index**, the first two pages list the organic functional groups, inorganic reagents, and general structural classes of reagents covered in the Encyclopedia with the page where the category can be found. The remainder of the Structural Class Index links the reactant type or functional group in each general reagent class with the volume (bold print) and page where each entry can be found in the Encyclopedia.

The **Subject Index** links the chemical reagents, named reactions, and functional groups with the volume in bold letters and page in the Encyclopedia where the entry can be found. The chemical names of reagents with their own entries are in bold print.

ORGANIZATION

The compounds are arranged alphabetically throughout the Encyclopedia. Most entries contain three lines consisting of the main entry, a description of its use, and the chemistry involved. Closely related compounds such as ether or ester derivatives (methyl, ethyl, etc.) are covered under one heading, usually the more developed reagent. Bold italics are used within an article to indicate any reagents that have their own entries in the Encyclopedia. The first literature reference in each entry is a review of the reagent. A list of related reagents is given at the end of each article. The abbreviations used for journal references use only the first letter of each word in the journal title. The abbreviations can be found on the front endpages of each volume and in section B.16.b of this book.

B.16.b.   JOURNAL ABBREVIATIONS FOR THE ENCYCLOPEDIA OF REAGENTS FOR ORGANIC SYNTHESIS

A

ABC	Agric. Biol. Chem.
AC(R)	Ann. Chim. (Rome)
ACR	Acc. Chem. Res.
ACS	Acta Chem. Scand.
AF	Arzneim.-Forsch.
AG	Angew. Chem.
AG(E)	Angew. Chem., Int. Ed. Engl
AJC	Aust. J. Chem.
AK	Ark. Kemi
ANY	Ann. N. Y. Acad. Sci.
AP	Arch. Pharm. (Weinheim, Ger.)

B

B	Biochemistry
BAU	Bull. Acad. Sci. USSR, Div. Chem. Sci.
BBA	Biochim. Biophys. Acta
BCJ	Bull. Chem. Soc. Jpn.
BJ	Biochem. J.MR
BML	Bioorg. Med. Chem. Lett.
BSB	Bull. Soc. Chim. Belg.
BSF(2)	Bull. Soc. Chim. Fr., Part 2

C

C	Chimia
CA	Chem. Abstr.
CB	Ber. Dtsch. Chem. Ges./Chem. Ber.
CC	Chem. Commun./J.. Chem. Soc.,
CCC	Collect. Czech. Chem. Commun.
CED	J. Chem. Eng. Data
CI(L)	Chem. Ind. (London)
CJC	Can. J. Chem.
CL	Chem. Lett.
COS	Comprehensive Organic Synthesis
CPB	Chem. Pharm. Bull.
CR(C)	C.R. Hebd. Seances Acad. Sci., Ser. C
CRV	Chem. Rev.
CS	Chem. Scr.
CSR	Chem. Soc. Rev.
CZ	Chem.-Ztg.

D

DOK	Dokl. Akad. Nauk SSSR

E

E	Experientia

F

FES	Farmaco Ed. Sci.
FF	Fieser&Fieser

G

G	Gazz. Chim. Ital.

H

H	Heterocycles
HC	Heteroatom Chem.
HCA	Helv. Chim. Acta.

I

IC	Inorg. Chem.
ICA	Inorg. Chim. Acta
IJC(B)	Indian J. Chem., Sect. B
IJS(B)	Int. J. Sulfur Chem., Part B
IZV	Izv. Akad. Nauk SSSR, Ser. Khim.

J

JACS	J. Am.. Chem. Soc.
JBC	J. Biol. Chem.
JCP	J. Chem. Phys.
JCR(M)	J. Chem. Res. (M)
JCR(S)	J. Chem. Res. ((S))
JCS	J.Chem. Soc
JCS(C)	J. Chem. Soc.
JCS(D)	J. Chem. Soc., Dalton Trans.
JCS(F)	J. Chem. Soc., Faraday Trans.
JCS(P1)	J. Chem. Soc., Perkin Trans. 1
JCS(P2)	J. Chem. Soc., Perkin Trans. 2
JFC	J. Fluorine Chem.
JGU	J. Gen. Chem. USSR (Engl. Transl)
JHC	J. Heterocycl. Chem.
JIC	J. Indian Chem. Soc.
JMC	J. Med. Chem.
JMR	J. Magn. Reson.
JOC	J. Org. Chem.
JOM	J. Organomet. Chem.
JOU	J. Org. Chem. USSR (Engl. Transl.)
JPOC	J. Phys. Org. Chem.
JPP	J. Photochem. Photobiol
JPR	J. Prakt. Chem.
JPS	J. Pharm. Sci.

K

KGS	Khim. Geterotsikl. Soedin.

L

LA	Justus Liebigs Ann. Chem./Liebigs Ann. Chem.

M

M	Monatsh. Chem.
MOC	Methoden Org. Chem. (Houben-Weyl)
MRC	Magn. Reson. Chem.

N

N	Naturwissenschaften
NJC	Nouv. J. Chim.
NKK	Nippon Kagaku Kaishi

O

OM	Organometallics
OMR	Org. Magn. Reson.
OPP	Org. Prep. Proced. Int.
OR	Org. React.
OS	Org. Synth.
OSC	Org. Synth., Coll. Vol.

P

P	Phytochemistry
PAC	Pure Appl. Chem.
PIA(A)	Proc. Indian Acad. Sci., Sect. A
PNA	Proc. Natl. Acad. Sci. USA
PS	Phosphorus Sulfur/Phosphorus/Sulfur Silicon

Q

QR	Q. Rev., Chem. Soc.

R

RCR	Russ. Chem. Rev. (Engl. Transl.)
RRC	Rev. Roum. Chem.
RTC	Recl. Trav. Chim. Pays-Bas

S

S	Synthesis
SC	Synth. Commun.
SL	Synlett

T

T	Tetrahedron
TA	Tetrahedron: Asymmetry
TL	Tetrahedron Lett.

U

| UKZ | Ukr. Khim. Zh. (Russ. Ed.) |

Y

| YZ | Yakugaku Zasshi |

Z

ZC	Z. Chem.
ZN(B)	Z. Naturforsch., Teil B
ZOB	Zh.Obshch. Khim.
ZOR	Zh. Org. Khim.

Encyclopedia of Reagents for Organic Synthesis (1995), L. A. Paquette (ed.). Copyright John Wiley & Sons Limited. Reproduced with permission.

B.17.

Formation of C–C Bonds (Mathieu):

QD262.M345

- **functional group transformations**
- **synthetic conversions**

B.17.a. OVERVIEW

The volumes focus on those organic reactions that can be utilized to prepare single or multiple carbon–carbon bonds for the construction of the carbon skeleton of molecules.

EASY ACCESS

— To find a desired transformation, check the detailed Table of Contents in the front of each volume.
— The Index of the Table of Contents (B.17.b) is an alphabetical listing of all the entries from the Table of Contents from each volume.

ORGANIZATION

The treatise is divided into three separate volumes with each volume using various sequence rules to organize the reactions covered. The volumes and reaction classifications are:

I. *Introduction of a Functional Carbon Atom*
    sequence rules:
    - Increasing oxidation state of the created functions;
    - sequential order: halogen, oxygen, sulfur, nitrogen, and other heteroatoms entering the reaction;
    - hybridization sp^3, sp^2, sp of the carbon atoms and of their substituents involved in the reaction.

II. *Introduction of a Carbon Chain or an Aromatic Ring*
    sequence rules:
    - substitution reactions precede addition reactions
    - the sequence of substitution reactions is given by the nature of the leaving group in the order: halogen, oxygen, sulfur, nitrogen, phosphorus.

    – Formation of single bonds between:
       – sp^3 carbon atoms
       – sp^3 and aromatic sp^2
       – sp^3 and vinylic sp^2
       – sp^3 and sp
       – aromatic sp^2
       – aromatic sp^2 and vinylic sp^2
       – aromatic sp^2 and sp
       – vinylic sp^2
       – vinylic sp^2 and sp
       – sp carbon atoms
    – Formation of double bonds

III. Introduction of an α-Functional Carbon Chain
    the reactions are classified with respect to the hybridization of the
    attacked carbon atom in the following order:
       – activated carbon atoms
       – aliphatic organometallics
       – olefins undergoing addition reactions
       – aromatic rings
       – aromatic organometallics
       – olefins undergoing substitution reactions
       – vinylic organometallics
       – 1-alkynes

(Reprinted by permission of Thieme New York, Inc.)

B.17.b.   INDEX TO THE TABLE OF CONTENTS OF FORMATION OF C–C BONDS

A
acylation, **IIIF**
acylation, thio-, **IIIG**
acyloxyalkylidenation, α-, **IIIF**
acyloxymethylation, **IB**
acyloxyvinylation, **IIIF**
akylidenation, α-amido, **IIID**
akylidenation, α-alkoxyl, **IIIF**
aliphatics, α-haloalkylation of, **IIIA**
alkoxycarbonylation, **IJ**
alkoxylakylidenation, α-, **IIIF**
alkoxylalkylation, **IIIB**
alkoxylation, gem-α-di-, **IIIF**

alkoxymethylation , **IB**
alkoxymethylenation, **IF**
alkylation, α-alkoxyl, **IIIB**
alkylation, α-amino, **IIID**
alkylation, dihalo-, gem-, **IIIE**
alkylation, dithio, **IIIG**
alkylation, α-hydroxyl, **IIIB**
alkylation, imino-, **IIIH**
alkylation, α-thio, **IIIC**
alkylation of 1-alkynes, **IID**
alkylation of activated carbon atoms, **IIA**
alkylation of aliphatic organometallics, **IIA**
alkylation of aromatic organometallics, **IIB**
alkylation of aromatic rings, **IIB**
alkylation of enamines, **IIA**
alkylation of olefins, **IIC**
alkylation of organboranes, **IIA**
alkylation of vinylic organometallics, **IIC**
alkylidenation, acyloxy, α-, **IIIF**
alkylidenation, amino-, **IIIH**
alkylidenation, thio-, **IIIG**
alkylidenation of diazoalkanes, **IIK**
alkylidenation of enamines, **IIK**
alkylidenation of enol ethers, **IIK**
alkylidenation of gem-organometallics, **IIK**
alkylidenation of methylene groups, **IIK**
alkylidenation of ylides, **IIK**
alkylthiomethylenation, **IG**
alkynes, alkylation of, **IID**
alkynes, alkynylation of, **IIJ**
alkynes, arylation of, **IIG**
alkynes, vinylation of, **II/I**
alkynylation of aliphatic organometallics, **IID**
alkynylation of 1-alkynes, **IIJ**
alkynylation of aromatic organometallics, **IIG**
alkynylation via organoboranes, **IID**
alkynylation of vinylic organometallics, **II/I**
amidination, **IL**
amidoalkylation, **IIID**
amidomethylation, **ID**
aminoalkylation, **IIID**
aminoalkylidenation, **IIIH**
aminomethylation, **ID**

aminomethylation, di-, **IH**
aminomethylenation, **IH**
aromatic rings, alkylation of, **IIB**
aromatic rings, arylation of, **IIE**
aromatic rings, vinylation of, **IIF**
aromatics, α-haloalkylation of, **IIIA**
arylation of activated carbon atoms, **IIB**
arylation of aliphatic organometallics, **IIB**
arylation of 1-alkynes, **IIG**
arylation of enamines, **IIB**
arylation of enol ethers, **IIB**
arylation of olefins, **IIB**
arylation of vinylic organometallics, **IIF**

C
carbamoylation, **IL**
carbamoylation, thio-, **IL**
carboxylation, **IJ**
carbonylation, alkoxy-, **IJ**
carbonylation, halo-, **IJ**
carboxylation, thio-, **IK**
cyanation, **IL**

D
dialkoxylation, gem-α-, **IIIF**
dialkoxymethylation, **IF**
dialkylthiomethylation, **IG**
diaminomethylation, **IH**
diazoalkanes, alkylidenation of, **IIK**
diazomethylation, **ID**
diazonium salts, **IIB**
dihaloalkylation, **IIIE**
dihalomethylenation, **I/I**
disulfonylmethylenation, **IK**
dithioalkylation, **IIIG**
dithiomethylenation, **IK**
dihalomethylation, **IE**

E
enamines, alkylation of, **IIA**
enamines, alkylidenation of, **IIK**
enamines, arylation of, **IIB**

enol ethers, alkylidenation of, **IIK**
enol ethers, arylation of, **IIB**

F
formylation, **IF**

H
haloalkoxymethylation, **IF**
haloalkylation, di-, gem-, **IIIE**
haloalkylation of aliphatics, **IIIA**
haloalkylation of aromatics, **IIIA**
halocarbonylation, **IJ**
halomethylation, **IA**
halomethylation, tri-, **I/I**
halomethylenation, **IE**
hydroxylalkylation, **IIIB**
hydroxymethylation, **IB**
hydroxymethylenation, **IF**

I
iminoalkylation, **IIIH**
iminomethylation, **IH**
immoniomethylation, **I**

M
methylation, acyloxy-, **IB**
methylation, alkoxy-, **IB**
methylation, amido-, **ID**
methylation, amino, **ID**
methylation, amino, di-, **IH**
methylation, diamino, **IH**
methylation, diazo-, **ID**
methylation, dihalo-, **IE**
methylation, halo-, **IA**
methylation, halo-alkoxy-, **IF**
methylation, dialkylthio-, **IG**
methylation, dialkoxy, **IF**
methylation, hydroxy-, **IB**
methylation, immonio-, **IH**
methylation, nitro-, **ID**
methylation, trinitro-, **IL**
methylation, thio, tri-, **IK**

methylation, halo, tri-, **I/I**
methylation, thio, dialkyl-, **IG**
methylation, imino-, **IH**
methylation, sulfamoyl-, **IC**
methylation, sulfinyl-, **IC**
methylation, sulfo-, **IC**
methylation, sulfonyl-, **IC**
methylation, thio-, **IC**
methylation, trihalo-, **I/I**
methylation, trithio-, **IK**
methylenation, alkoxy, **IF**
methylenation, alkylthio-, **IG**
methylenation, amino-, **IH**
methylenation, dihalo-, **I/I**
methylenation, disulfonyl-, **IK**
methylenation, dithio-, **IK**
methylenation, halo, **IE**
methylenation, halo, di-, **I/I**
methylenation, hydroxy-, **IF**
methylenation, nitro-, **IH**
methylenation, sulfonyl, di-, **IK**
methylenation, thio, alkyl-, **IG**
methylenation, thio, di-, **IK**

N
nitromethylation, **ID**
nitromethylation, tri-, **IL**
nitromethylenation, **IH**

O
olefins, alkylation of, **IIC**
olefins, arylation of, **IIB**
olefins, vinylation of, **IIH**
organoboranes, alkylation of, **IIA**
organoboranes, alkynylation, **IID**
organoboranes, vinylation of, **IIC**
organoboranes, vinylation of, **IIH**
organometallics, aliphatic, alkylation of, **IIA**
organometallics, aliphatic, alkynylation, **IID**
organometallics, aliphatic, vinylation of, **IIC**
organometallics, alkylidenation of, gem- **IIK**

organometallics, aromatic, alkylation of, **IIB**
organometallics, aromatic alkynylation of **IIG**
organometallics, aromatic, arylation of, **IIE**
organometallics, aromatic, vinylation of, **IIF**
organometallics, arylation of, **IIB**
organometallics, vinylic, alkylation of, **IIC**
organometallics, vinylic alkynylation of, **II/I**
organometallics, vinylic, arylation of, **IIF**
organometallics, vinylic, vinylation of, **IIH**
oxyvinylation, acyl-, **IIIF**

S
sulfamoylmethylation, **IC**
sulfinylmethylation, **IC**
sulfomethylation, **IC**
sulfonylmethylation, **IC**

T
thioacylation, **IIIG**
thioalkyklation, **IIIC**
thioalkylation, di-, **IIIG**
thioalkylidenation, **IIIG**
thiocarbamoylation, **IL**
thiocarboxylation, **IK**
thiomethylation, **IC**
thiomethylation, dialkyl-, **IG**
thiomethylation, tri-, **IK**
thiomethylenation, alkyl-, **IG**
trihalomethylation, **I/I**
trinitromethylation, **IL**
trithiomethylation, **IK**

V
vinylation, acyloxy-, **IIIF**
vinylation of activated carbon atoms, **IIC**
vinylation of aliphatic organometallics, **IIC**
vinylation of 1-alkynes, **II/I**
vinylation of aromatic organometallics, **IIF**
vinylation of aromatic rings, **IIF**
vinylation of olefins, **IIH**
vinylation of organoboranes, **IIC**

vinylation of organoboranes, **IIH**
vinylation of vinylic organometallics, **IIH**

Y
ylides, alkylidenation of, **IIK**

(Reprinted by permission of Thieme New York, Inc.)

B.18.

Houben-Weyl Methoden der Organischen Chemie:

QD258.M5

- **preparation and analysis of *classes* of compounds**
- **rearrangements**
- **example experimentals**

B.18.a.   OVERVIEW

The treatise covers the reaction chemistry, synthesis, and analysis of various classes of organic compounds. Representative experimentals and pertinent literature references accompany each review.

EASY ACCESS

— The book bindings list the class of compounds found in that particular volume.
— Volume 16/1 is the Index of Experimental Procedures and 16/2 is the Index of Substance Classes.
— All subject headings can be found in the Index of Houben-Weyl Table of Contents, Section B.18.b of this book. For more specific information, refer to the author and subject index in each volume.

INDEXING

Each volume has its own author and subject index. The Author Index links the names of referenced authors and coauthors with the page in that volume where they are cited. The Subject Index links the names of the chemical substances with the page in which they appear. The compounds are named by the Beilstein system and some trivial and tradenames are included. Page numbers that appear in bold print indicate that an experimental procedure to make the compound is given on that page. See the Table of Contents for collective names of compound classes.

The **Index of Experimental Procedures** in Volume 16/1 has an alphabetical index of compound types and the volume and page where experimentals for preparing the class of substances can be found. The **Formula Index** of procedures to prepare individual compounds is listed in the back of this volume. The index links the molecular formula with the name of the compound and the volume in bold type, and page where the procedure can be found. Formulas are arranged by citing first the number of carbons, then hydrogens, and finally the remaining elements in alphabetical order by element symbol. If carbon is not present, the elements are arranged in alphabetical order. Carboxylic acid salts are listed under the parent acid and basic metal salts under the metal atom.

The **Index of Classes of Compounds** in Volume 16/2 is divided into two subsections: 2a. Open chain compounds including alkaloids and sugars; and 2b. Cyclic and spiro compounds and steroids. Subsection 16/2a consists of a short Table of Contents that lists the general classes of compounds with the section and page to find the substance class in the Fine Table of Contents. The Fine Table of Contents further divides each class by substitution or derivatization and gives the page number in the index where the sub-class can be found. The Index of Classes of Compounds gives the compound name, volume in bold type, and the page where the substance is located.

Subsection 16/2b consists of a **Table of Contents and Substance Index**. The Table of Contents lists cyclic compounds according to ring size, followed by those containing heteroatoms, bicyclic compounds in order to ring size combinations, and finally steroids. The heteroatoms follow the order: O, S, Se, Te, N, P, As, Sb, and then the metals by periodic table group number. Saturated cyclic compounds are listed before their unsaturated forms. The cyclic substance class is linked to the page where the compounds appear in the index. The Index of Classes of Compounds gives the compound name, structure, volume in bold type, and the page where the substance is located.

The **Named Reactions Index** is located at the end of volume 16/2b. It is an alphabetical listing of named reactions with volumes and page numbers where the reaction is discussed in Houben-Weyl.

The **Index to the Complete E Series** will consist of ten subsections in Volume E 22 and should begin publication in 1998–99. The sections consist of:

E 23a General Compound Classes (alphabetically) Formula Index
   Individual Compounds (C1–C11)
E 23b Formula Index Individual Compounds (>C11)
E 23c Substance Index Aliphatic Compounds I Carbonyl Compounds
   (CO_2 to Acetals)
E 23d Substance Index Aliphatic Compounds II Alcohols to Alkanes
E 23e Substance Index Cyclic Compounds I 3- and 4-Membered
   Monocyclic Compounds
E 23f Substance Index Cyclic Compounds II 5-Membered
   Monocyclic Compounds

E 23g Substance Index Cyclic Compounds III n-Membered
Monocyclic Compounds (n > 5)
E 23h Substance Index Cyclic Compounds IV Bicyclic Compounds I
E 23i Substance Index Cyclic Compounds V Bicyclic Compounds II
E 23j Substance Index Cyclic Compounds VI Tricyclic and
Polycyclic Compounds

INTERNET

A complete listing of all the volumes with ISBN numbers is available at:

http://www.thieme.com/adb.htm

ORGANIZATION

Houben-Weyl is a 16-volume set divided into seven major sections and
23 supplemental volumes ("E Series").
The major sections are:

I.   General Laboratory Practice (Vol. 1)
II.   Analytical Methods (Vol. 2)
III.   Physical Methods of Investigation (Vol. 3)
IV.   General Chemical Methods (Vol. 4)
V.   Chemistry of Substance Classes (Vols. 5–13)
VI.   Macromolecular Substances (Vol. 14)
VII.   Synthesis of Peptides (Vol. 15)
VIII. Index (Vol. 16)

B.18.b.   HOUBEN-WEYL INDEX TO THE TABLE OF CONTENTS

A
acetals 6/3, 6/4, E14
acetals, hemi-, 6/4
acetals, unsaturated 7/4
acetylenes, poly-, cyclic 5/1d
acid halides E5
acidity, C-H 13/1
acids 8, E5
acids, amino 11/2
acids, thio 9
acridine E7b
alcohols 6/1a, 6/1b
aldehydes 7/1, E3

aldehydes, thio 9
alkanes 5/1a
alkanes, cyclic 5/1a
alkenes 5/1b
alkenes, conjugated 5/1c
alkenes, vicinal dihalo 7/4
alkenes, with alkynes 5/1d
alkoxides 6/2
alkynes 5/2a
alkynes, unsaturated 5/1d
alkynes, with alkenes 5/1d
aluminium, organometallics 13/4
amides 8, E5
amides, polymeric 14/2
amides, thio 9
amines E16a, E16d
amine oxides E16a1
amines, preparation 11/1
amines, transformations 11/2
amino acids 11/2, E5
ammonium salts, quaternary 11/2, E16a2
anhydrides E5
antimony derivatives 6/2
arenes 5/2b
arsenic derivatives 6/2
azepines E9d
azetidine 11/2
azines 10/2
aziridines 11/2
azo compounds 10/3, E16d
azodicarboxylates 10/2
azodioxy compounds, aromatic 10/3
azoxy compounds, aromatic 10/3

B

barium, organometallics 13/2a
benzofuran E6b$_1$/b$_2$
benzothiophene E6b$_1$/b$_2$
beryllium, organometallics 13/2a
biphosphine 12/1
borate esters 6/2
bromides 5/4

C

cadmium, organometallics 13/2a
calcium, organometallics 13/2a
calicenes E17d
carbanions E19d
carbazole E6a
carbenes E19b
carbocations E19c
carbodiimide E4
carbon dioxide E4
carbon disulfide E4
carbonates, polymeric 14/2
carbonyls, polymeric 14/2
carboxylation 8
carboxylic acids 8, E5
cellulose 14/2
cesium, organometallics 13/1
chlorides 5/3, 5/4
cinnolines E9a
copper, organometallics 13/1
cyanines 5/1d
cyclazines E9d
cycloalkanes 5/1a
cyclobutadienes E17f
cyclobutanes 4/4, E17e
cyclobutenes E17f
cyclobutenediones E17f
cycloheptatriene 5/1d
cyclooctane, unsaturated 5/1d
cyclopropane 4/3, E17a, b
cyclopropanes, reactions E17c
cycloproparenes E17d
cyclopropene 4/3, E17d
cyclopropenones E17d
cyclopropenylium salts E17d

D

decarboxylation 8
diazenes, aliphatic 10/2
diazenes, aliphatic N-oxides 10/2
diazines E9a, E9b
diazinodiazines E9c
diazirines 10/2; 10/4

diazo compounds, aliphatic 10/4
diazonium salts, aromatic 10/3, E16a2
dibenzofuran E6a
dibenzothiophene E6a
Diels-Alder reaction 5/1c
dienes 5/1b
dienes, conjugated 5/1c
dioxanes 6/4
dioxazines E9c
dioxins E9a
diselenides 9
disulfides 9
dithiins E9a
dithio acids E5
dithio esters E5
dithiols 9

E
ene-ynes 5/1d
epoxides 6/3
epoxides, thio 9
esters 8, E5
esters, ortho 6/3
esters, polymeric 14/2
esters, thio 9
ethers 6/3
ethers, bridged 6/4
ethers, cyclic, five-membered 6/3
ethers, cyclic, six-membered 6/4
ethers, cyclic, seven or larger 6/4
ethers, oxonium ions 6/3
ethylenesulfide 9

F
fluorides 5/3, E10
fluorine E10
formazane 10/3
furan E6a
furans, tetrahydro 6/3

G
gallium, organometallics 13/4
gold, organometallics 13/1

Grignard reagents 13/2a
guanidine 8

H
halogenophosphines 12/1 p127
halogens 5
haloamines E16a2
hetarenes E9
heteroaromatics, 7-ring E9d
heterocycles, 6-ring E9a
hydrazines, 10/2, E16a2
hydrazones,10/2
hydrazones, aromatic 10/3
hydroxylamines 10/1, E5, E16a1
hypobromous acid 6/2
hypochlorous acid 6/2
hypoiodous acid 6/2

I
index 16/1,2, E9e, E23
indium, organometallics 13/4
iodides 5/4
indole $E6b_1/b_2$
indolizine $E6b_1/b_2$
isocyanate E4
isocyanide E5
isonitriles 8
isoquinoline E7a
isothiocyanates 9, E4

K
ketene 7/4
ketones, by decomposition 7/2b
ketones, preparation 7/2a
ketones, reactions 7/2b
ketones, thio 9

L
lactams 11/2, E16b
lactones 6/2
lithium, organometallics 13/1

M
macrocycles E20
macromolecular, analysis 14/2
macromolecular materials 14/1
macromolecular, structure 14/2
magnesium, organometallics 13/2a
mercaptans 9
mercury, organometallics 13/2b
morpholine 6/4

N
nitrate esters 8, 9
nitrenes E16c
nitrile oxides E5
nitriles 8, E4, E5
nitriles, N-oxide 10/3
nitrite esters 8, 9
nitro, 10/1, E16d
nitro amines E16a2
nitrones 10/4
nitroso, 10/1, E16a2
nitroso amines E16a2
nitrous acid esters E16c
nitrous acid E16c

O
organocuprates 13/1
organolithiums 13/1
organometallics 13/1, E18
organotellurium E12b
orthoesters 6/3, E5
oxathiazines E9c
oxathiins E9a
oxazinium salts E9a
oxaziridine 10/4
oxepins E9d
oxetanes 6/3
oximes 10/4
oxirans 6/3
oxirans, thio 9
oxonium ions 6/3

P

peptide bonds, preparation 15/2
peptide, nomenclature 15/1
peptide purification 15/2
peptides E22
peptides, synthesis 15/1
peptidomemetics E22
perchlorate derivatives 6/2
peroxides 8, E13
phenols 6/1c
phenoxazines E9a
phosgene 8, E4
phosphates E1, 2
phosphines 12/1
phosphinic acid 12/1
phosphinous acid 12/1
phosphonium salts 12/1
phosphonous derivatives 12/1,2
phosphoranes, halo- 12/1
phosphoranes, oxy- 12/1
phosphoric acid 12/2
phosphorous E1,2
phthalazines E9a
phthalocyanies E9d
polyene, open chain 5/1d
polymerization,14/1
polymers 14/1, 14/2
polymers, amides 14/2
polymers, analysis 14/2
polymers, carbonates 14/2
polymers, carbonyl 14/2
polymers, conversions 14/2
polymers, esters 14/2
polymers, heteroatom monomers 14/2
polymers, monomers 14/1,2
polymers urethanes 14/2 p57
polysulfides 9
porphyrins E9d
potassium, organometallics 13/1
pteridines E9c
purines E9b
pyrans, 6/4
pyrazines E9b

pyridazines E9a
pyridine E7b
pyrimidines E9b
pyrrole E6a

Q

quinazolines E9b
quinoline E7a

R

radicals E19a
rubber, conversions 14/2
rubidium, organometallics 13/1

S

selenenic acids 9
selenides 9
seleninic acids 9
selenium derivatives 9
silanes 6/2
silver, organometallics 13/1
sodium, organometallics 13/1
starch 14/2
stereoselective synthesis E21
strontium, organometallics 13/2a
sulfate esters 6/2
sulfides 9
sulfonamides 11/2
sulfonate esters 6/2
sulfonic acid derivatives 6/2
sulfonimides 11/2
sulfoxides, 9
sulfur compounds E11
sulphenamides 9
sulphenate esters 9
sulphenic acid 9
sulphenyl halides 9
sulphimide 9
sulphinic acids, 9
sulphonamides 9
sulphonate esters, aromatic 9
sulphonic acids, anhydrides, 9

sulphonic acids, 9
sulphonium salts 9
sulphonyl azides, aromatic 9
sulphonyl chlorides 9
sulphonyl halides, aromatic 9
sulphonyl hydrazines 9
sulphoxides 9
sulphoximide 9
synthesis, stereoselective E21

T
tellerium, organo- E12b
tellurenyl compounds 9
tellurium compounds E12b
tellurium derivatives 9
tellurols 9
tetrazines 10/2, E9c
thallium, organometallics 13/4
thiafulvenes E17d
thiazines E9a, E9c
thiepins E9d
thio amides E5
thio esters E5
thioacids 9
thioaldehydes 9
thioamides 9
thiocarbonates 9
thiocyanates 9
thioesters 9
thioethers 9
thioketene 7/4
thioketones 9
thiols 9
thiomorpholine 6/4
thionaphthoquinone 7/4
thiophene E6a
thiophenol 9
thiophosgene E4
thioureas 9
triazines, aliphatic 10/2, E16a2
triazines, aromatic 10/3

U
urethanes, polymeric 14/2

W
Wittig reagents 12/1

X
xanthene E7b

Z
zinc, organometallics 13/2a

(Reprinted by permission of Thieme New York, Inc.)

B.19.

Internet Resources

Articles

"The World-Wide Web as a Chemical Information Tool," Peter Murray-Rust, Henry S. Rzepa and Benjamin J. Whitaker, Chemical Society Reviews, 1997, 1.

Books

Bachrach, Stephen M. editor *The Internet: a Guide for Chemists* Washington, DC: American Chemical Society, 1996. (QD8.3 .I57 1996)

Clement, Gail P. *Science and Technology on the Internet: an Instructional Guide* Berkeley, CA: Library Solutions Press, 1995. (Q179.96 .C54 1995)

Dix, James *The Houghton Mifflin Guide to the Internet for Chemistry*, Boston: Houghton Mifflin, 1996.

Lee, C.C., *Chemical Guide to the Internet* Rockville, Md.: Government Institutes, c 1996. (QD31.2 .L42 1996)

Ridley, Damon D., *On-Line Searching: A Scientist's Perspective*, New York: Wiley, 1996.

Thomas, Brian J. *The Internet for Scientists and Engineers: Online Tools and Resources* Bellingham, WA, USA: SPIE Optical Engineering Press, 1995. (TK5105.875 .I57 T48 1995)

Internet Resources

ChemCenter; A Service of the American Chemical Society

> http://www.ChemCenter.org

Chemical Information Sources from Indiana University: CIS-IU (Gary Wiggins)

> http://www.indiana.edu/~cheminfo/

Chemistry: Yahoo

> http://www.yahoo.com/Science/Chemistry

ChemWeb.com

> http://ChemWeb.com

InfoSurf: Chemistry

   http://www.library.ucsb.edu/subj/chemistr.html

WebChemistry

   http://www.latrobe.edu.au/www/webchem/

World-Wide Web Virtual Library: Chemistry

   http://www.chem.ucla.edu/chempointers.html

B.20.

Journals

The internet addresses given in the table correspond to web pages where the journal's table of contents can be accesed at no cost. Access to full-text journals of important chemistry publications is available through the following:

ChemPort   www.chemport.org
ChemWeb   www.chemweb.com
Ebsco Online   www.ebsco.com/ess/electron/bropage1.asp
Electronic Journals Navigatore   services.blackwell.co.uk/ejn
Information Quest   www.informationquest.com
SwetsNet   www.swetsnet.com

Journal	Call No.	Internet Address
Accounts of Chemical Research	QD1.A275	pubs.acs.org/journals/achre4/index. html
Advances in Carbohydrate Chemistry	QD321.A2	www.chemistry.ohio-state.edu/~lowary/advances.html

Journal	Call No.	Internet Address
Aldrichimica Acta	QD1.A3529a	www.sigma.sial.com/aldrich/acta/index.htm
Angewandte Chemie International Edition	QD1.A67	www.vchgroup.de/contents/jc_2002/index.html
Applied Organometallic Chemistry	QD410.A66	sunsite.nus.sg/wiley-text/che/che35.html
Australian Journal of Chemistry	QD1.A85	www.isb.csiro.au/cis/journals/ajc/index.html
Beilstein Home Page	QD251.B4	www.beilstein.com/index.html
Bulletin of the Chemical Society of Japan	QD1.T63	www.syp.toppan.co.jp:8082/bcsjstart.html
Bulletin de la Society Chimique de France	QD1.S4	www.elsevier.nl/locate/bulchimf
Bulletin of the Czech Chemical Society		www.uochb.cas.cz/bulletin.html
Canadian Journal of Chemistry	QD1.C285	www.cisti.nrc.ca/cisti/journals/rjchem.html
Carbohydrate Research	QD321.C3	www.elsevier.nl/locate/carabsonline
Chemical Society of Japan, Bulletin	QD1.T63	www.syp.toppan.co.jp:8082/bcsjstart.html
Chemical Abstracts Service	QD1.A51	info.cas.org/welcome.html
Chemical Web Marketing Technology		www.chemconnect.com/cwmt.shtml
Chemical Communications	QD1.C525	www.rsc.org/is/journals/current/chemcomm/cccpub.htm
Chemical Reviews	QD1.C45	pubs.acs.org/journals/chreay/index.html
Chemical Society Reviews	QD1.C558	www.rsc.org/is/journals/current/chsocrev/csrpub.htm
Chemical Processing	TP1.C367	www.chemicalprocessing.com

Chemicke Listy	QD1.C54	www.uochb.cas.cz/ChemListy/chemlisty.html
Chemische Berichte	QD1.D4	www.vchgroup.de/contents/jc_2005/index.html
Chemistry Letters	QD1.C744	wwwsoc.nacsis.ac.jp/csj/journals/chem-lett/index-e.html
Chemistry	QD1.C74	ci.mond.org
Chemistry of Heterocyclic Compounds	QD400.K513	www.osi.lanet.lv/hgs/hgs.html
Collection of Czechoslovak Chemical Comm	QD1.C8	cccc.uochb.cas.cz
Current Organic Chemistry	QD241.C87	www.bscipubl.demon.co.uk/coc/index.html
European Journal of Organic Chemistry	QD241.E97	www.wiley-vch.de/vch/journals/2046/index.html
Faraday Discussions	QD1.F32	chemistry.rsc.org/rsc/fadpub.htm
Flavour & Fragrance Journal		sunsite.nus.sg/wiley-text/che/che41.html
Food Chemistry		www.elsevier.nl/locate/foodchem
Food & Chemical Toxicology		www.elsevier.nl/locate/foodchemtox
Fullerenes Alert		www.elsevier.nl/locate/fullalert
Helvetica Chimica Acta	QD1.H4	www.vchgroup.de/vhca/journals/2217.html
Heterocycles	QD399.H63	www.elsevier.nl:80/locate/estoc/03855414
Journal of Agricultural and Food Chemistry	S583.J6	pubs.acs.org/journals/jafcau/index.html
Journal of Fluorine Chemistry	QD181.F1 J66	www.elsevier.nl:80/locate/estoc/00221139
Journal of Labelled Compounds	QD466.J59	sunsite.nus.sg/wiley-text/che/che50.html
Journal of Medicinal Chemistry	RS402.J65	pubs.acs.org/journals/jmcmar/index.html
Journal of Natural Products	QH1.L94	pubs.acs.org/journals/jnprdf/index.html

Journal	Call No.	Internet Address
Journal of Organic Chemistry	QD241.J6	pubs.acs.org/journals/joceah/index.html
Journal of Organic Chemistry Electronic Supporting Information		pubs.acs.org/supmat/joceah
Journal of Organometallic Chemistry	QD411.J65	www.elsevier.nl:80/locate/estoc/0022328X
Journal of Physical Organic Chemistry		sunsite.nus.sg/wiley-text/che/che54.html
Journal of the American Chemical Society	QD1.A5	pubs.acs.org/journals/jacsat/index.html
Journal of the Chemical Society - Perkin Trans 1	QD1.C55	www.rsc.org/is/journals/current/perkin1/p1ppub.htm
Journal of the Chemical Society - Perkin Trans 2	QD1.C552	www.rsc.org/is/journals/current/perkin2/p2ppub.htm
Journal of the Chemical Society - Faraday Discus	QD1.F32	www.rsc.org/is/journals/current/faraday/fadpub.htm
Journal of the Chemical Society - Dalton Trans	QD1.C53	www.rsc.org/is/journals/current/dalton/dappub.htm
Liebigs Annalen	QD1.L7	www.vchgroup.de/contents/jc_2046/
Medicinal Chemistry Research		www.birkhauser.com/journals/mcr/mcr.html
Molecules; Journal of organic synthesis		science.springer.de/molec/molecule.htm
Monatshefte Chemie/Chemical Monthly	QD1.M7	link.springer.de/link/service/journals/00706/index.htm

Journal	Call Number	URL
Natural Product Report	QD415.A1 N378	www.rsc.org/is/journals/current/npr/nprpub.htm
New Journal of Chemistry	QD1.N848	www.rsc.org/is/journals/current/newjchem/njc.htm
Organic Process Research & Development	TP155.7 O7	pubs.acs.org/journals/oprdfk/index.html
Organometallics	QD410.O733	pubs.acs.org/journals/orgnd7/index.html
Phytochemistry	QK861.P45	www.elsevier.nl/locate/phytochem/
Polyhedron	QD1.P64	www.elsevier.nl:80/locate/inca/218
Pure and Applied Chemistry	QD1.I88152	www.blacksci.co.uk/products/journals/pac.htm
Recueil des Travaux Chimique des Pays-Bas	QD1.R3	www.elsevier.nl:80/inca/publications/store/5/2/1/2/1/9/
Russian Chemical Reviews	QD1.U713	rcr.ioc.ac.ru/rcr.html
Steroids	QP752.S7 S75	www.elsevier.nl:80/locate/inca/525022
Synlett	QD262.S96	www.thieme.com/chemistry/Aeae.htm
Synthesis	QD262.S93	www.thieme.com/chemistry/Aebe.htm
Synthetic Communications	QD262.S935	www.dekker.com/cgi-bin/webdbc/md/detail.htx?d_cat_id=0039-7911
Talanta	QD71.T3	www.elsevier.nl/locate/talanta/
Tetrahedron Information System		oxford.elsevier.com/tis
Tetrahedron News		www.elsevier.nl/locate/inca/31080
Tetrahedron Asymmetry	QD481.T43	sgdsec.dp.ox.ac.uk/
Tetrahedron	QD241.T47	www.elsevier.nl/locate/tetrahedron/
Tetrahedron Letters	QD241.T475	www.elsevier.nl/locate/tetletts/
Topics in Current Chemistry	QD1.F58	gopher://trick.ntp.springer.de:70/11/chemistry/tcc/

B.21.

Kirk-Othmer Encyclopedia of Chemical Technology:

TP9.E685

- general subjects in chemical technology
- properties and manufacture of compounds
- uses of various substances
- industrial processes

The entries consist of applied chemical science and industrial technology, its methods and materials, as well as the latest scientific advances in every branch of the useful arts of chemistry and related general subjects.

EASY ACCESS

— Consult the General Index of the Index to Volumes 1 to 25 and Supplement, fourth edition.

INDEXING

The Index to Volumes 1 to 25 and Supplement, fourth edition, is divided into four sections: Contents of the Encyclopedia, Contributors to the Encyclopedia, General Index, and Chemical Abstracts Registry Numbers. The **Contents of the Encyclopedia** lists of the general subjects covered in each volume. The entries are alphabetical within each volume, and the corresponding page number follows the subject title. Important subheadings are listed in italic.

The **Contributors to the Encyclopedia** links the names of Encyclopedia authors together with their institution or company (in italic) and entry title. In the case of co-authors, each author is listed separately.

The **General Index** lists encyclopedia entries, chemicals, general subjects, etc. with the volume number in italic followed by a colon then the page number. Article titles are indicated by a boldface page number, and the CAS Registry Number is given in square brackets following the name of an indexed compound.

The **Chemical Abstracts Service Registry Numbers** Index lists the CA Registry Number of every indexed compound with its corresponding chemical name.

ORGANIZATION

Articles are arranged alphabetically through out Kirk-Othmer. The treatment of a compound will be found either under its own name, or under a group of substances (for example, ethyl acetate under Esters, organic), or as a derivative under a parent compound (for example, ethyl acrylate under Acrylic acid and derivatives). The properties and manufacture of any substance are given in one article, which makes cross-reference to one or more articles where the uses of that substance are described. The article entries for each edition of the Encyclopedia may not be included in subsequent printings or covered by previous editions; therefore, a search of all editions of Kirk-Othmer may be necessary to find a particular subject. The Encyclopedia printing dates are: 4th edition: 1991–1998; 3rd edition: 1978–1984; 2nd edition: 1963–1972; and, 1st edition: 1949–1956.

B.22.

Lange's Handbook of Chemistry:

- **compound name and structure**
- **physical and chemical properties**
- **conversion and mathematical tables**

B.22.a. OVERVIEW

The Handbook contains descriptive properties of organic and inorganic compounds as well as conversion factors and common laboratory techniques. Particular attention is paid to natural products, alkaloids, and glucosides.

EASY ACCESS

— Organic compounds are listed alphabetically by their IUPAC name in Section 7.
— Consult the Index after Section 11 in the back of the Handbook to locate the desired table.
— Use the Index to the Table of Contents, Section B.17.b, to determine if the information you require is covered in Lange's Handbook.

INDEXING

The **Index** (after Section 11) is located in the back of the Handbook and indexes pertinent properties, general concepts, classes of substances, and some common substances.

The **Empirical Formula Index of Organic Compounds** (Section 7–14) lists the molecular formula of chemical substances, with their Table identification number.

In **Physical Constants of Organic Compounds** (Section 7–15), the organic compounds in are listed alphabetically according to their IUPAC senior prefix.

ORGANIZATION

The Handbook is divided into 11 sections numbered with the section number in bold face followed by the page number (**3**–133 is Section 3, page 133). The titles of the sections and subsections are indexed in B.22.b.

B.22.b.   INDEX TO THE SECTIONS IN LANGE'S HANDBOOK

A

Abbreviations and Standard Letter Symbols, Section 2

Abbreviations, Mathematical, Section 2

Absorbance for Percent Absorption, Values of, Section 2

Absorbance-Transmittance Conversion Table, Section 2

Absorption Coefficients, Mass, Section 8

Absorption Coefficients, Mass, for Ka_1 Lines and W La_1 Line, Section 8

Absorption Edges in Angstroms, Wavelengths of, Section 8

Absorption and Emission Lines 1900 to 9000 Å, Section 8

Absorption Energies, X-ray, in keV, Section 8

Absorption Spectroscopy, Section 8

Absorption, Values of Absorbance for, Section 2

Acid-Base Solvent, Properties of, Section 5

Acid-Base Titrations, Aqueous, and Colorimetric Determination of pH, Indicators for, Section 5

Acidities of Various Compounds in Nonaqueous Solvents, Section 5

Acidity in 50wt% MeOH-Water, Standard Ref. Values for the Measure of, Section 5

Acidity in Heavy Water, Standard Ref. Values pD, for the Measurement of, Section 5

Acidity in Other Solvent Media, Section 5

Acids and Bases Laboratory Solutions, Concentrations of, Section 11

Acids, Mineral, in Aqueous Solutions at $25\,^\circ C$, H_o for, Section 5

Acids, Organic, Binary Azeotropes Containing, Section 10

Acids, Organic, Solubilities in Water at Various Temperatures, Section 10

Activity Coefficients, Section 5

Activity Coefficients of Ions in Water at $25\,^\circ C$, Section 5

Activity Coefficients, Ionic, at Higher Ionic Strengths at $25\,^\circ C$, Section 5

Activity Standards, Ion, Section 5

Affinities, Electron, Section 3

Air, Moist, Density of, Section 10

Air, Properties of Combustible Mixtures in, Section 11

Air, Specific Gravity at Various Temperature, Section 10

Air, Specific Gravity Corrections for the Buoyant Effect of, Section 10

Air, Saturated, Mass of Water Vapor in, Section 10

Alcohols and Phenols, Retained Trivial Names of with Structures, Section 7

Alcohols and Water, Ternary Azeotropes Containing, Section 10

Alcohols, Binary Azeotropes Containing, Section 10
Alkaloids, Physical Constants of, Section 7
Alkane Carbons, Estimation of Chemical Shift of, Section 8
Alkanes, Straight-Chain, Names of, Section 7
Alkyl Chemical Shifts, Effect of Sustituent Groups on, Section 8
Alloys and Metals, Thermal Conductivity of, Section 10
Alphabet, Greek, Section 2
Aluminum Oxide, Heat Capacity Standards, Section 9
Amino Acids and Polypeptides, Hormonal, Section 7
Ammonia, Liquid NH_3, Vapor Pressure of, Section 10
Analytical Chemistry, Section 5
Analyzing Crystals for X-ray Spectroscopy, Section 8
Angle, Trigonometric Functions, Section 1
Anilines, Hammett Sigma Constants for para Substituents in, Section 3
Anions, Inorganic, Anodic Depolarization Potentials of, Section 6
Anions on Dowex Resins, Selectivity Coefficients E X/Cl for, Section 5
Anodic Depolarization Potentials of Inorganic Anions, Section 6
Antifreeze Solutions, Aqueous, Compositions of, Section 10
Antimony Electrode, Section 5
Approximate Effective Ionic Radii in Aqueous Solutions at 25°C,
    Section 5
Approximate Potential Ranges in Nonaqueous Solvents, Section 5
Aqueous Antifreeze Solutions, Compositions of, Section 10
Aqueous Glycerol Solutions, Section 10
Aqueous Solutions, Approximate Effective Ionic Radii in, Section 5
Aqueous Solutions, Coefficients of Compressibility of, Section 10
Aqueous Solutions, Coefficients of Cubical Expansion for, Section 10
Aqueous Solutions of H_2SO_4, NaOH, and $CaCl_2$, Aqueous Tensions
    of, Section 10
Aqueous Solutions of H_2SO_4, NaOH, and $CaCl_2$, Relative Humidities
    of, Section 10
Aqueous Sucrose Solutions, Section 10
Aqueous Tensions of Aqueous Solutions of H_2SO_4, NaOH, and $CaCl_2$,
    Section 10
Areas and Volumes, Surface, Section 1
Areas between Abscissa Values $-z$ and $+z$ of the Distribution Curve,
    Section 1
Atmospheric Data, Section 2
Atomic and Group Refractions, Section 10
Atomic and Molecular Structure, Section 3
Atomic Electron Affinities, Section 3
Atoms and Ions, Radii of, Section 3
Azeotropes, Binary, Containing Alcohols, Section 10

Azeotropes, Binary, Containing Organic Acids, Section 10
Azeotropes, Binary, Containing Water, Section 10
Azeotropes, Ternary, Containing Water and Alcohols, Section 10
Azeotropic Mixtures, Section 10

B

β Filters for Common Target Elements, Section 8
Barometer Temperature Correction, Section 2
Barometer to Sea Level, Reduction of, Section 2
Barometric Corrections, Section 2
Barometric Latitude-Gravity Table, Section 2
Barometric Pressure Corrections for the Hydrogen Electrode, Section 5
Barometric Pressures at Temperatures, Boiling Points of Water, Section 10
Barometry and Barometric Corrections, Section 2
Bases and Acids, Laboratory Solutions, Concentrations of, Section 11
Baths, Nonaqueous Cooling, Section 10
Batteries, Section 6
Benzene, Monosubstituted, Chemical Shifts in, Section 8
Benzenes, Substituted, Carbon-13 Chemical Shifts in, Section 8
Binary Azeotropes Containing Alcohols, Section 10
Binary Azeotropes Containing Organic Acids, Section 10
Binary Azeotropes Containing Water, Section 10
Biological Media, Standards for pH Measurement of, Section 5
Blood and Biological Media, Standards for pH Measurement of, Section 5
Boiling Points, Section 10
Boiling Point at Various Pressures, Correction of, A.S.T.M. Method, Section 10
Boiling Point at Various Pressures, Correction of, Dreisbach Method, Section 10
Boiling Point (Ebullioscopic Constants), Molecular Elevation of, Section 10
Boiling Points of Water at Pressures, Section 10
Boiling Points of Water, Barometric Pressures at Temperatures, Section 10
Boiling Points of Organic Compounds, Calculation of, Section 10
Boiling Points, Organic Solvents Arranged by, Section 10
Bond Energies, Section 3
Bonds, Hybrid, Spatial Orientation of, Section 3
British Standards Method, pH Values of Buffers by, Section 5
Btu, Mineral-free, and Fixed Carbon, Formulas for Calculating, Section 11

Buffer Solutions, Composition and pH Values of, Section 5
Buffer Solutions for Control Prposes, pH Values of, Section 5
Buffer Solutions in Alcohol-Water Solvents at 25 °C, pH* Values for, Section 5
Buffer Solutions other than standards, Section 5
Buffer Solutions, Reference pH, National Bureau of Standards (U.S.), Section 5
Buffer Solutions, Standard pH, Compositions of, NBS (U.S.), Section 5
Buffer Solutions, Standard Reference pH, Section 5
Buffers by the British Standards Method, pH Values of, Section 5
Buoyant Effect of the Air, Specific Gravity Corrections, Section 10
Burets, Volumetric, Tolerances for, Section 2

C

Calcium Chloride, Aqueous Tensions of Aqueous Solutions of, Section 10
Calcium Chloride, Relative Humidities of Aqueous Solutions of, Section 10
Calculation of Boiling Points of Organic Compounds, Section 10
Calculation of Concentrations of Species Present at a Given pH, Section 5
Calculation of the Approximate pH Values of Solutions, Section 5
Calibrating Conductivity Vessels, Standard Solutions for, Section 6
Capacity, Permissible Deviations from, Section 2
Carbon, Fixed, and Mineral-free Btu, Formulas for Calculating, Section 11
Carbon-13 Chemical Shifts, Section 8
Carbon-13 Chemical Shifts in Substituted Benzenes, Section 8
Carboxylic Acids, Names of, Section 7
Cells, Voltaic, and Batteries, Section 6
Characteristic Groups Cited Only as Prefixes in Nomenclature, Section 7
Characteristic Groups for Substitutive Nomenclature, Section 7
Chemical Properties of Natural Resins, Section 7
Chemical Shift for Protons of $-CH_2-$ and $>CH-$ Groups, Estimation of, Section 8
Chemical Shift of Alkane Carbons (Relative to TMS), Estimation of, Section 8
Chemical Shift of Carbon Attached to a Double Bond, Estimation of, Section 8
Chemical Shift of Proton Attached to a Double Bond, Estimation of, Section 8
Chemical Shifts, Effect of Sustituent Groups on Alkyl, Section 8

Chemical Shifts, Carbon-13, Section 8
Chemical Shifts, Carbon-13, in Substituted Benzenes, Section 8
Chemical Shifts in Monosubstituted Benzene, Section 8
Chemical Shifts, Proton, Section 8
Chemical Symbols and Terminology, Section 2
Chromatography, Column, Section 5
Coal from Analyses, Heat of Combustion of, Section 11
Coefficient of Linear Expansion, Section 10
Coefficients, Activity, Section 5
Coefficients of Cubical Expansion for Aqueous Solutions, Section 10
Coefficients of Cubical Expansion for Various Liquids, Section 10
Coefficients of Compressibility of Aqueous Solutions, Section 10
Coefficients of Compressibility of Various Liquids, Section 10
Coefficients, Selectivity, E M/H/n for Ions on Cross-Linked Dowex 50,
    Section 5
Coefficients, Selectivity, E X/Cl for Some Anions on Dowex Resins,
    Section 5
Colorimetric Determination of pH, Indicators for, Section 5
Column Chromatography, Section 5
Column Efficiency, Section 5
Combustible Mixtures in Air, Properties of, Section 11
Commercial Organic Materials, Section 7
Common Acid-Base Solvent, Properties of, Section 5
Common Electrode Reactions at 25 °C, Overpotentials for, Section 6
Complexes, Metal, Formation Constants of, Section 5
Composition and pH Values of Buffer Solutions, Section 5
Compositions of Aqueous Antifreeze Solutions, Section 10
Compositions of Standard pH Buffer Solutions, NBS (U.S.), Section 5
Compressibility and Thermal Expansion, Section 10
Compressibility of Aqueous Solutions, Coefficients of, Section 10
Compressibility of Various Liquids, Coefficients of, Section 10
Compressibility of Water, Section 10
Computing HNP Values from pK_a Values, Constants for, Section 5
Concentration Units, Relation of, Section 11
Concentrations of Commonly Used Acids and Bases Laboratory Solu-
    tions, Section 11
Concentrations of Species Present at a Given pH, Calculation of,
    Section 5
Conductances, Limiting Equivalent Ionic, in Aqueous Solutions at
    25 °C, Section 6
Conductivities of Electrolytes in Aqueous. Solutions at 18 °C, Section 6
Conductivity, Electrical, of Various Pure Liquids, Section 6
Conductivity of Very Pure Water at Various Temperatures, Section 6

Conductivity Vessels, Standard Solutions for Calibrating, Section 6
Constant Humidity, Solutions for Maintaining, Section 10
Constants, Cryoscropic, Section 10
Constants, Equilibrium, Section 5
Constants for Computing HNP Values from pKa Values, Section 5
Constants, Formation, of Metal Complexes, Section 5
Constants, Mathematical, Section 1
Constants of Fats, Oils, and Waxes, Section 7
Constants of the Debye-Huckel Equation from 0 to 100°C, Section 5
Constants, Physical, Section 2
Conversion Factors, Section 2
Conversion of Specific Gravity to Density at Any Temperatures, Section 10
Conversion of Thermometer Scales, Section 2
Conversion of Weighings in Air to Weighings in Vacuo, Section 2
Conversion Table, Hydrometer, Section 2
Conversion Table, Temperature, Section 2
Conversion Table, Viscosity, Section
Conversion Table, Wavenumber/Wavelength, Section 2
CONVERSION TABLES, Section 2.
Conversions of Density-Specific Gravity, Section 10
Cooling Baths, Nonaqueous, Section 10
Correction for Emergent Stem of Liquid-in-Glass Thermometers, Section 11
Correction of Boiling Point at Various Pressures, A.S.T.M. Method, Section 10
Correction of Boiling Point at Various Pressures, Dreisbach Method, Section 10
Corrections, Specific Gravity, for the Buoyant Effect of the Air, Section 10
Coupling Constants, Proton Spin, Section 8
Critical Phenomena, Section 9
Critical Properties, Section 9
Critical X-ray Absorption Energies in keV, Section 8
Cross-Linked Dowex 50, Selectivity Coefficients E M/H/n for Ions on, Section 5
Cryoscropic Constants, Section 10
Crystal Lattice Types, Section 3
Crystal Structure, Section 3
Crystals, Analyzing for X-ray Spectroscopy, Section 8
Cubical Coefficients of Thermal Expansion, Section 2
Cubical Expansion for Aqueous Solutions, Coefficients of, Section 10
Cubical Expansion for Various Liquids, Coefficients of, Section 10

Cubical Expansion of Solids, Section 10
Cumulative Formation Constants for Metal Complexes with Inorganic Ligands, Section 5
Cumulative Formation Constants for Metal Complexes with Organic Ligands, Section 5

D

Debye-Huckel Equation from 0 to 100°C, Constants of, Section 5
Density and Specific Gravity, Section 10
Density at Any Temperature, Conversion of Specific Gravity to, Section 10
Density of Moist Air, Section 10
Density of Mercury and Water, Section 10
Density-Specific Gravity, Conversions of, Section 10
Depolarization Potentials, Anodic, of Inorganic Anions, Section 6
Derivatives and Differentiation, Section 1
Deuterated Solvents, Positions of Residual Protons in, Section 8
Dew Point and Humidity from Wet and Dry Bulb Readings, Section 10
Dielectric Constants, Section 10
Dielectric Constant of Selected Inorganic Substances, Section 10
Dielectric Constant of Water, Section 10
Differentiation, Section 1
Dipole Moment of Selected Inorganic Substances, Section 10
Dipole Moments, Section 10
Distillation, Theoretical Plates, Section 5
Distribution Curve, Ordinates and Areas, Abscissa Values $-z$ and $+z$, Section 1
Distribution, F, Section 1
Distribution, x^2, Percentiles, Section 1
Double Bond, Estimation of Chemical Shift of Carbon Attached to, Section 8
Double Bond, Estimation of Chemical Shift of Proton Attached to, Section 8
Dowex 50, Cross-Linked, Selectivity Coefficients E M/H/n for Ions on, Section 5
Dowex Resins, Selectivity Coefficients E X/Cl for Some Anions on, Section 5
Dry and Wet Bulb Readings, Humidity and Dew Point from, Section 10
Drying Agents, Section 10
Drying and Humidification, Section 10
Drying Oils, Glyceride Content of, Section 7

E

Ebullioscopic Constants, Section 10

Edges, Absorption, in Angstroms, Wavelengths of, Section 8

Efficiency, Column, Section 5

Electrical Conductivity of Various Pure Liquids, Section 6

Electrochemistry, Section 6

Electrode, Antimony, Section 5

Electrode, Hydrogen, Section 5

Electrode, Hydrogen, Barometric Pressure Corrections for, Section 5

Electrode Potentials, Single, Section 6

Electrode, Quinhydrone, Section 5

Electrode, Quinhydrone, Ref. Electrodes, Standard Potentials of, Section 5

Electrode Reactions at 25°C, Overpotentials for, Section 6

Electrodes, Ref., in Volts for Water-Org. Solvent Mixtures, Potentials of Section 5

Electrolytes in Aqueous Solutions at 18°C, Equivalent Conductivities of, Section 6

Electrometric Measurement of pH, Section 5

Electron Affinities, Atomic, Section 3

Electronegativities of the Elements, Section 3

Electronic Emission and Absorption Spectroscopy, Section 8

Elements and Inorganic Compounds, Section 9

Elements and Their Compounds at 25°C, Potentials of, Section 6

Elements, Electronegativities, Section 3

Elements, Ionization Potentials, Section 3

Elements, Nuclear Properties of, Section 3

Elements, Physical Properties of, Section 3

Elements, Sensitive Lines of, Section 8

Elements, Target, ß Filters for, Section 8

Emergent Stem of Liquid-in-Glass Thermometers, Correction for, Section 11

Emission and Absorption Lines 1900 to 9000 Å, Section 8

Emission, Electronic, and Absorption Spectroscopy, Section 8

Emission Energies, X-Ray, in keV and Filters, Section 8

Emission Spectra, X-ray, Wavelengths of, in Angstroms, Section 8

Empirical Formula Index for Organic Compounds, Section 7

Energies, Bond, Section 3

Energies, Critical X-ray Absorption, in keV, Section 8

Energies, X-Ray Emission, in keV and Filters, Section 8

Enthalpies of Formation of Elements and Compounds, Section 9

Entropies of Elements and Compounds, Section 9

Equations and Equivalents for Titrimetric (Volumetric) Analysis, Section 5

Equations for Gases, Section 11

Equilibrium Constants, Section 5

Equilibrium Constants in Aqueous Solutions, Temperature Dependence of, Section 5

Equivalent Conductivities of Electrolytes in Aqueous Solutions at 18 °C, Section 6

Equivalent Ionic Conductances, Limiting, in Aqueous Solutions at 25 °C, Section 6

Equivalents for Titrimetric (Volumetric) Analysis, Section 5

Errors of Indicators, Salt, Section 5

Estimation of Chemical Shift for Protons of $-CH_2-$ and $>CH$ - Groups, Section 8

Estimation of Chemical Shift of Alkane Carbons (Relative to TMS), Section 8

Estimation of Chemical Shift of Carbon Attached to a Double Bond, Section 8

Estimation of Chemical Shift of Proton Attached to a Double Bond, Section 8

Estimation of Gibbs Free Energy, Section 9

Expansion, Cubical, for Aqueous Solutions, Coefficients of, Section 10

Expansion, Cubical, for Various Liquids, Coefficients of, Section 10

Expansion in Series, Section 1

Expansion, Linear, Coefficient of, Section 10

Expansion, Thermal, and Compressibility, Section 10

Expansion, Thermal, Cubical Coefficients of, Section 2

F

F Distribution, Section 1

Factors for Simplified Computation of Volume, Section 2

Factors, Gravimetric, Section 5

Fats, Oils, and Waxes, Constants of, Section 7

Filters and X-Ray Emission Energies in keV, Section 8

Filters, ß, or Common Target Elements, Section 8

Fixed Carbon and Mineral-free Btu, Formulas for Calculating, Section 11

Fluorescent Indicators, Section 5

Formation Constants of Metal Complexes, Section 5

Formation Constants for Metal Complexes with Inorganic Ligands, Section 5

Formation Constants for Metal Complexes with Organic Ligands, Section 5

Formula Index for Organic Compounds, Section 7
Formulas for Calculating Mineral-free Btu and Fixed Carbon, Section 11
Formulas of Synthetic Rubbers, Section 7
Formulas of Thermoplastic and Thermosetting Materials, Section 7
Formulas and Names of Organic Radicals, Section 7
Free Energy of Component Groups, Section 9
Freezing Mixtures, Section 10
Functional Class Names Used in Radiofunctional Nomenclature, Section 7
Fundamental Physical Constants, Section 2
Fused Polycyclic Hydrocarbons, Section 7
Fusion Names, Trivial Names of Heterocyclic Systems Suitable for Use in, Section 7
Fusion Names, Trivial Names for Heterocyclic Systems that are *not* suitable for Use in, Section 7

G

Gas at T, Molar Equivalent of One Liter of, Section 2
Gases and Vapors, Thermal Conductivity of, Section 10
Gases in Water, Solubility of, Section 10
Gases, Physical Chemistry Equations for, Section 11
Gases, Van der Waals' Constants for, Section 11
Gastrointestinal Tract, Hormones of, Section 7
General Information and Conversion Tables, Section 2.
Gibbs (Free) Energies of Formation of Elements and Compounds, Section 9
Gibbs Free Energy, Estimation of, Section 9
Glucosides, Physical Constants of, Section 7
Glyceride Content of Drying Oils, Section 7
Glycerol Solutions, Aqueous, Section 10
Gravimetric Factors, Section 5
Greek Alphabet, Section 2
Group Refractions, Section 10
Groups Cited Only as Prefixes in Nomenclature, Section 7
Groups for Substitutive Nomenclature, Section 7
Groups, Free Energy of, Section 9

H

Half-Reactions at 25 °C, Potentials of, Section 6
Half-Wave Potentials of Inorganic Ions at 25 °C, Section 6
Half-Wave Potentials of Organic Compounds at 25 °C, Section 6
Hammett and Taft Equations, Section 3

Hammett and Taft Substituent Constants, Section 3

Hammett Equation, pK_a^0 and p Values for Section 3

Hammett Sigma Constants for para Substituents in Anilines, Section 3

Hammett Sigma Constants for para Substituents in Phenols, Section 3

Hammett Sigma Constants for para Substituents in Pyridines, Section 3

Heat Capacities of Elements and Compounds, Section 9

Heat Capacity Standards: Water, Mercury, Aluminum Oxide, Section 9

Heat of Combustion of Coal from Analyses, Section 11

Heat of Sublimation of the Elements and Inorganic Compounds, Section 9

Heat of Vaporization of the Elements and Inorganic Compounds, Section 9

Heats of Melting of the Elements and Inorganic Compounds, Section 9

Heavy Water, Standard Ref. Values pD, for the Measurement of Acidity in, Section 5

Heterocyclic Systems, Specialist Nomenclature for, Section 7

Heterocyclic Systems, Suffixes for Specialist Nomenclature of, Section 7

Heterocyclic Systems Suitable for Use in Fusion Names, Trivial Names of, Section 7

Heterocyclic Systems that are **not** suitable for Use in Fusion Names, Trivial Names for, Section 7

HNP Values from pK_a Values, Constants for Computing, Section 5

H_o for Mineral Acids in Aqueous Solutions at 25 °C, Section 5

Hormonal Amino Acids and Polypeptides, Section 7

Hormonal Steroids, Section 7

Hormones of the Gastrointestinal Tract, Section 7

Hormones, Properties of, Section 7

Hormones, Steroid, Section 7

Humidification and Drying, Section 10

Humidity and Dew Point from Wet and Dry Bulb Readings, Section 10

Humidity, Solutions for Maintaining Constant, Section 10

Hybrid Bonds, Spatial Orientation of, Section 3

Hydrocarbons, Fused Polycyclic, Section 7

Hydrogen Electrode, Section 5

Hydrogen Electrode, Barometric Pressure Corrections for, Section 5

Hydrometer Conversion Table, Section 2

Hydrometers, Section 10

I

Ice, Vapor Pressure of, in Millimeters of Mercury, Section 10

Incompletely Deuterated Solvents, Positions of Residual Protons in, Section 8

Index Follows Section 11

Index, Empirical Formula, for Organic Compounds, Section 7

Indicators, Fluorescent, Section 5

Indicators for Aqueous. Acid-Base Titrations and Colorimetric Determination of pH, Section 5

Indicators, Mixed, Section 5

Indicators, Oxidation-Reduction, Section 6

Indicators, Salt Errors of, Section 5

Individual Activity Coefficients of Ions in Water at 25 °C, Section 5

Individual Ionic Activity Coefficients at Higher Ionic Strengths at 25 °C, Section 5

Inorganic Anions, Anodic Depolarization Potentials of, Section 6

Inorganic Chemistry, Section 4

Inorganic Compounds, Section 9

Inorganic Compounds, Nomenclature of, Section 4

Inorganic Compounds, Physical Constants of, Section 4

Inorganic Compounds, Solubilities in Water at Various Temperatures, Section 10

Inorganic Compounds, Vapor Pressures of, Section 10

Inorganic Ions at 25 °C, Half-Wave Potentials of, Section 6

Inorganic Ligands, Metal Complexes with Formation Constants for, Section 5

Inorganic Materials in Water at 25 °C, Proton-transfer Reactions of, Section 5

Inorganic Substances, Dielectric Constant of, Section 10

Inorganic Substances, Dipole Moment of, Section 10

Inorganic Substances, Surface Tension of, Section 10

Inorganic Substances, Viscosity of, Section 10

Integrals, Section 1

Interplanar Spacings, Section 8

Interplanar Spacings for Ka_1 Radiation, d vs 2θ, Section 8

Ion Activity Standards, Section 5

Ion-Exchange Parameters, Section 5

Ionic Activity Coefficients at Higher Ionic Strengths at 25 °C, Section 5

Ionic Conductances, Limiting Equivalent, in Aqueous Solutions at 25 °C, Section 6

Ionic Production Constant of Water, Section 5

Ionic Radii in Aqueous Solutions at 25 °C, Section 5

Ionic Strengths, Higher, at 25 °C, Individual Ionic Activity Coefficients, Section 5

Ionization Potentials of Molecular and Radical Species, Section 3

Ionization Potentials of the Elements, Section 3

Ions, Inorganic, at 25 °C, Half-Wave Potentials of, Section 6

Ions in Water at 25 °C, Individual Activity Coefficients of, Section 5
Ions on Cross-Linked Dowex 50, Selectivity Coefficients E M/H/n for, Section 5
Ions, Radii of, Section 3

K
Ka$_1$ Lines and W La$_1$ Line, Mass Absorption Coefficients for, Section 8
Ka$_1$ Radiation, d vs 20, Interplanar Spacings for, Section 8
Kopp's Rule, Section 9

L
Latitude-Gravity Table, Barometric, Section 2
Lattice Types, Crystal, Section 3
Letter Symbols, Section 2
Limiting Equivalent Ionic Conductances in Aqueous Solutions at 25 °C, Section 6
Linear Expansion, Coefficient of, Section 10
Lines, Sensitive, of the Elements, Section 8
Liquids and Solutions, Thermal Conductivity of, Section 10
Liquids, Coefficients of Compressibility of, Section 10
Liquids, Coefficients of Cubical Expansion for, Section 10
Liquids, Electrical Conductivity of, Section 6
Logarithms: Properties and Uses, Section 1

M
Magnetic Resonance, Nuclear, Section 8
Maintaining Constant Humidity, Solutions for, Section 10
Mass Absorption Coefficients, Section 8
Mass Absorption Coefficients for Ka$_1$ Lines and W La$_1$ Line, Section 8
Mass of Water Vapor in Saturated Air, Section 10
Materials, Commercial Organic, Section 7
Mathematical Symbols and Abbreviations, Section 2
Mathematical Tables, Section 1
MATHEMATICS, Section 1
Measure of Acidity in 50wt% MeOH-Water, Standard Ref. Values for, Section 5
Measurement of Acidity in Heavy Water, Standard Ref. Values pD, Section 5
Measurement of pH, Section 5
Measurement of pH, Electrometric, Section 5
Measurement of Temperature, Section 11
Melting Points, Section 10

Mercury, and Aluminum Oxide, Heat Capacity Standards, Section 9
Mercury and Water, Density of, Section 10
Mercury, Vapor Pressure of, Section 10
Metal Complexes, Formation Constants of, Section 5
Metal Complexes with Inorganic Ligands, Cumulative Formation Constants for, Section 5
Metal Complexes with Organic Ligands, Cumulative Formation Constants for, Section 5
Metal Salts, Solubilities in Water at Various Temperatures, Section 10
Metals and Alloys, Thermal Conductivity of, Section 10
Methine Protons, Estimation of Chemical Shift for, Section 8
Methylene Protons, Estimation of Chemical Shift for, Section 8
Mineral Acids in Aqueous Solutions at 25 °C, H_o for, Section 5
Mineral-free Btu and Fixed Carbon, Formulas for Calculating, Section 11
Mineral Names, Section 4
Miscellaneous, Section 11
Mixed Indicators, Section 5
Mixtures, Azeotropic, Section 10
Mixtures, Freezing, Section 10
Mixtures in Air, Combustible, Properties of, Section 11
Moist Air, Density of, Section 10
Molar Equivalent of One Liter of Gas at T, Section 2
Molecular and Radical Species, Ionization Potentials of, Section 3
Molecular Elevation of the Boiling Point (Ebullioscopic Constants), Section 10
Molecular Structure, Section 3

N

Names and Formulas of Organic Radicals, Section 7
Names of Some Carboxylic Acids, Section 7
Names of Straight-Chain Alkanes, Section 7
Names Used in Radiofunctional Nomenclature, Section 7
Naming Multiples and Submultiples of Units, Prefixes for, Section 2
National Bureau of Standards (U.S.) Reference pH Buffer Solutions, Section 5
Natural and Synthetic Rubbers, Properties of, Section 7
Natural Resins, Physical and Chemical Properties of, Section 7
NMR, Section 8
NMR, Carbon-13 Chemical Shifts, Section 8
NMR, Carbon-13 Chemical Shifts in Substituted Benzenes, Section 8
NMR, Chemical Shifts in Monosubstituted Benzene, Section 8
NMR, Effect of Substituent Groups on Alkyl Chemical Shifts, Section 8

NMR, Estimation of Chemical Shift of Alkane Carbons (Relative to TMS), Section 8

NMR, Estimation of Chemical Shift of Carbon Attached to a Double Bond, Section 8

NMR, Estimation of Chemical Shift of Proton Attached to a Double Bond, Section 8

NMR, Estimation of Chemical Shift for Protons of $-CH_2-$ and $>CH-$ Groups, Section 8

NMR, Positions of Residual Protons in Incompletely Deuterated Solvents, Section 8

NMR, Proton Chemical Shifts, Section 8

NMR, Proton Spin Coupling Constants, Section 8

Nomenclature, Section 7

Nomenclature, Characteristic Groups Cited Only as Prefixes, Section 7

Nomenclature, Characteristic Groups for, Section 7

Nomenclature, for Heterocyclic Systems, Section 7

Nomenclature of Heterocyclic Systems, Suffixes for, Section 7

Nomenclature of Inorganic Compounds, Section 4

Nomenclature, Organic Chemical, Short Glossary of, Section 7

Nomenclature, Radiofunctiona, Functional Class Names Used in, Section 7

Nonaqueous Cooling Baths, Section 10

Nonaqueous Solvents, Acidities of Various Compounds in, Section 5

Nonaqueous Solvents, Approximate Potential Ranges in, Section 5

Nonaqueous Solvents, pK_a Values for Proton-Transfer Reaction in, Section 5

Nuclear Magnetic Resonance, Section 8

Nuclear Properties of the Elements, Section 3

Nuclides, Table of, Section 3

Numerical Prefixes, Section 2

O

Oils, and Waxes, Constants of, Section 7

Oils, Drying, Glyceride Content of, Section 7

Ordinates between Abscissa Values $-z$ and $+z$ of the Distribution Curve, Section 1

Organic Acids, Binary Azeotropes Containing, Section 10

Organic Acids, Solubilities in Water at Various Temperatures, Section 10

Organic Chemical Nomenclature, Short Glossary of, Section 7

Organic Chemistry, Section 7

Organic Compounds, Section 9

Organic Compounds at 25 °C, Half-Wave Potentials of, Section 6

Organic Compounds, Calculation of Boiling Points of, Section 10
Organic Compounds, Empirical Formula Index for, Section 7
Organic Compounds, Physical Constants of, Section 7
Organic Compounds, Vapor Pressures of, Section 10
Organic Ligands, Metal Complexes with, Cumulative Formation Constants for, Section 5
Organic Materials, Commercial, Section 7
Organic Materials in Water at 25°C, Proton-Transfer Reactions of, Section 5
Organic Radicals, Names and Formulas of, Section 7
Organic Solvents Arranged by Boiling Points, Section 10
Organic Solvents, Physical Properties of, Section 10
Orientation, Spatial, of Common Hybrid Bonds, Section 3
Overpotentials for Common Electrode Reactions at 25°C, Section 6
Oxidation-Reduction Indicators, Section 6

P
p Values for Hammett Equation, Section 3
p Values for Taft Equation, Section 3
Parent Structures of Phosphorus-Containing Compounds, Section 7
pD, Standard Ref. Values, for the Measurement of Acidity in Heavy Water, Section 5
Percentiles of the x^2 Distribution, Section 1
Permissible Deviations from Nominal Capacity, Section 2
pH Buffer, Standard Reference Solutions, Section 5
pH Buffer Solutions, Standard, NBS (U.S.), Compositions of, Section 5
pH Buffer Solutions, Reference, National Bureau of Standards (U.S.), Section 5
pH, Calculation of Concentrations of Species Present at Section 5
pH, Electrometric Measurement of, Section 5
pH, Indicators for Aqueous Acid-Base Titrations and Colorimetric Determinetion of, Section 5
pH in Other Solvent Media, Section 5
pH, Measurement of, Section 5
pH Measurement of Blood and Biological Media, Standards for, Section 5
pH Values of Buffers by the British Standards Method, Section 5
pH Values of Buffer Solutions, Section 5
pH Values of Buffer Solutions for Control Purposes, Section 5
pH Values of Solutions, Calculation of, Section 5
pH* Values for Buffer Solutions in Alcohol-Water Solvents at 25°C, Section 5

Phenols and Alcohols, Retained Trivial Names of with Structures, Section 7
Phenols, Hammett Sigma Constants for para Substituents in, Section 3
Phosphorus-Containing Compounds, Parent Structures of, Section 7
Physical and Chemical Properties of Natural Resins, Section 7
Physical and Chemical Symbols and Terminology, Section 2
Physical Constants, Section 2
Physical Constants of Alkaloids, Section 7
Physical Constants of Glucosides, Section 7
Physical Constants of Inorganic Compounds, Section 4
Physical Constants of Organic Compounds, Section 7
Physical Properties, Section 10
Physical Properties of Pure Substances, Section 7
Physical Properties of Refrigerants, Section 10
Physical Properties of Selected Organic Solvents, Section 10
Physical Properties of the Elements, Section 3
Pipets, Volumetric, Tolerances for, Section 2
pK_a^0 and p Values for Hammett Equation, Section 3
pK_a^0 and p Values for Taft Equation, Section 3
pK_a Values, Constants for Computing HNP Values from, Section 5
pK_a Values for Proton-Transfer Reaction in Nonaqueous Solvents, Section 5
Polarography, Section 6
Polycyclic Hydrocarbons, Fused, Section 7
Polypeptides and Amino Acids, Hormonal, Section 7
Positions of Residual Protons in Incompletely Deuterated Solvents, Section 8
Potential Ranges in Nonaqueous Solvents, Section 5
Potentials, Anodic Depolarization, of Inorganic Anions, Section 6
Potentials, Half-Wave, of Inorganic Ions at 25 °C, Section 6
Potentials, Half-Wave, of Organic Compounds at 25 °C, Section 6
Potentials, Ionization, of the Elements, Section 3
Potentials, Ionization, of Molecular and Radical Species, Section 3
Potentials of Reference Electrodes in Volts as a Function of Temperatures, Section 5
Potentials of Ref. Electrodes in Volts for Water-Org. Solvent Mixtures, Section 5
Potentials of the Elements and Their compounds at 25°C, Section 6
Potentials of the Quinhydrone Electrode Ref. Electrodes, Section 5
Potentials of Selected Half-Reactions at 25 °C, Section 6
Potentials, Single Electrode, Section 6
Prefixes for Naming Multiples and Submultiples of Units, Section 2

Prefixes in Nomenclature, Characteristic Groups Cited Only as, Section 7
Pressure, Section 2
Pressure Corrections, Barometric, for the Hydrogen Electrode, Section 5
Pressure-Vapor Equations, Section 10
Pressures, Boiling Points of Water at, Section 10
Pressures, Correction of Boiling Point at, A.S.T.M. Method, Section 10
Pressures, Correction of Boiling Point at, Dreisbach Method, Section 10
Properties, Nuclear, of the Elements, Section 3
Properties of Combustible Mixtures in Air, Section 11
Properties of Common Acid-Base Solvent, Section 5
Properties of Common Steroids, Section 7
Properties of Hormones, Section 7
Properties of Natural and Synthetic Rubbers, Section 7
Properties of Thermoplastic and Thermosetting Materials, Section 7
Properties, Physical, of the Elements, Section 3
Proton Chemical Shifts, Section 8
Proton Spin Coupling Constants, Section 8
Proton-transfer Reactions, Section 5
Proton-transfer Reactions of Inorganic Materials in Water at 25°C, Section 5
Proton-Transfer Reaction in Nonaqueous Solvents, pK_a Values for, Section 5
Protons, Residual, in Incompletely Deuterated Solvents, Positions of, Section 8
Pure Substances, Physical Properties of, Section 7
Pyridines, Hammett Sigma Constants for para Substituents in, Section 3

Q
Quinhydrone Electrode, Section 5
Quinhydrone Electrode Ref. Electrodes, Standard Potentials of, Section 5

R
Radiation, Ka_1, for d vs 20, Interplanar Spacings, Section 8
Radical Species, Ionization Potentials of, Section 3
Radicals, Organic, Names and Formulas of, Section 7
Radii of Atoms and Ions, Section 3
Radiofunctional Nomenclature, Functional Class Names Used in, Section 7
Reactions at 25°C, Overpotentials for Common Electrode, Section 6
Reactions, Proton-transfer, Section 5

Reactions, Proton-transfer, of Inorganic Materials in Water at 25 °C, Section 5

Reactions, Proton-Transfer, in Nonaqueous Solvents, pK_a Values for, Section 5

Reactions, Proton-transfer, of organic Materials in Water at 25 °C, Section 5

Reduction of the Barometer to Sea Level, Section 2

Reference Electrodes in Volts as a Function of Temperatures, Potentials of, Section 5

Reference pH Buffer Solutions, Section 5

Reference pH Buffer Solutions, National Bureau of Standards (U.S.), Section 5

Refractions, Atomic and Group, Section 10

Refractive Index, Section 10

Refractive Index of Water, Section 10

Refrigerants, Physical Properties of, Section 10

Relation of Concentration Units, Section 11

Relative Humidities of Aqueous Solutions of H_2SO_4, NaOH, and $CaCl_2$, Section 10

Resins, Natural, Physical and Chemical Properties of, Section 7

Resolution, Section 5

Retained Trivial Names of Alcohols and Phenols with Structures, Section 7

Retention Behavior, Section 5

Rubbers, Synthetic, Section 7

Rubbers, Synthetic, Formulas of, Section 7

Rubbers, Natural and Synthetic, Properties of, Section 7

S

Salt Errors of Indicators, Section 5

Salts, Metal, Solubilities in Water at Various Temperatures, Section 10

Saturated Air, Mass of Water Vapor in, Section 10

Screens and Sieves, Section 11

Sea Level, Reduction of the Barometer to, Section 2

Selected List of Oxidation-Reduction Indicators, Section 6

Selectivity Coefficients E M/H/n for Ions on Cross-Linked Dowex 50, Section 5

Selectivity Coefficients E X/Cl for Some Anions on Dowex Resins, Section 5

Sensitive Lines of the Elements, Section 8

Separation Methods, Section 5

Series, Expansion in, Section 1

Short Glossary of Organic Chemical Nomenclature, Section 7

Sieve Series, U.S., Section 11
Sieves and Screens, Section 11
Sigma Constants for para Substituents in Anilines, Section 3
Sigma Constants for para Substituents in Phenols, Section 3
Sigma Constants for para Substituents in Pyridines, Section 3
Single Electrode Potentials, Section 6
Sodium Hydroxide, Aqueous Tensions of Aqueous Solutions of, Section 10
Sodium Hydroxide, Relative Humidities of Aqueous Solutions of, Section 10
Solids, Cubical Expansion of, Section 10
Solubilities, Section 10
Solubilities of Inorganic Compounds in Water at Various Temperatures, Section 10
Solubilities of Metal Salts in Water at Various Temperatures, Section 10
Solubilities of Organic Acids in Water at Various Temperatures, Section 10
Solubility of Gases in Water, Section 10
Solubility Products, Section 5
Solutions and Liquids, Thermal Conductivity of, Section 10
Solutions, Aqueous Antifreeze, Compositions of, Section 10
Solutions, Aqueous, Approximate Effective Ionic Radii in, Section 5
Solutions, Aqueous, at $25\,°C$, H_0 for Mineral Acids in, Section 5
Solutions, Aqueous, at $18\,°C$, Equivalent Conductivities of Electrolytes in, Section 6
Solutions, Aqueous, Coefficients of Cubical Expansion for, Section 10
Solutions, Aqueous, Coefficients of Compressibility, Section 10
Solutions, Aqueous, of H_2SO_4, NaOH, and $CaCl_2$, Relative Humidities of, Section 10
Solutions, Aqueous, of H_2SO_4, NaOH, and $CaCl_2$, Aqueous Tensions of, Section 10
Solutions, Aqueous, Temperature Dependence of Equilibrium Constants in, Section 5
Solutions, Buffer, Composition and pH Values of, Section 5
Solutions, Buffer, other than standards, Section 5
Solutions, Calculation of the Approximate pH Values of, Section 5
Solutions, Concentrations of Commonly Used Acids and Bases, Section 11
Solutions for Calibrating Conductivity Vessels, Section 6
Solutions for Maintaining Constant Humidity, Section 10
Solutions, Standard pH Buffer, Compositions of, NBS (U.S.), Section 5
Solutions, Standard Reference pH Buffer, Section 5
Solutions, Standard Stock, Section 11

Solutions, Standard Volumetric (Titrimetric), Section 5
Solvent, Acid-Base, Properties of, Section 5
Solvents, Nonaqueous, Acidities of Various Compounds in, Section 5
Solvents, Nonaqueous, Approximate Potential Ranges in, Section 5
Solvents, Nonaqueous, pK_a Values for Proton-Transfer Reaction in,
    Section 5
Solvents, Organic, Arranged by Boiling Points, Section 10
Solvents, Organic, Physical Properties of, Section 10
Some Common Spectroscopic Relationships, Section 8
Some Physical Chemistry Equations for Gases, Section 11
Spacings, Interplanar, Section 8
Spacings, Interplanar, for Ka_1 Radiation, d vs 20, Section 8
Spatial Orientation of Common Hybrid Bonds, Section 3
Specialist Nomenclature for Heterocyclic Systems, Section 7
Specific Heat of the Elements and Inorganic Compounds, Section 9
Specific Gravity and Density, Section 10
Specific Gravity Corrections for the Buoyant Effect of the Air,
    Section 10
Specific Gravity-Density, Conversions of, Section 10
Specific Gravity of Air at Various Temperature, Section 10
Specific Gravity to Density at Any Temperatures, Conversion of,
    Section 10
Spectroscopic Relationships, Section 8
Spectroscopy, Section 8
Standard Letter Symbols, Section 2
Standard pH Buffer Solutions, NBS (U.S.), Compositions of, Section 5
Standard Potentials of the Quinhydrone Electrode Ref. Electrodes,
    Section 5
Standard Ref. Values for the Measure of Acidity in 50wt% MeOH-
    Water, Section 5
Standard Ref. Values pD, for the Measurement of Acidity in Heavy
    Water, Section 5
Standard Reference pH Buffer Solutions, Section 5
Standard Solutions for Calibrating Conductivity Vessels, Section 6
Standard Stock Solutions, Section 11
Standard Volumetric (Titrimetric) Solutions, Section 5
Standards for pH Measurement of Blood and Biological Media,
    Section 5
Standards, Heat Capacity: Water, Mercury, Aluminum Oxide, Section 9
Standards, Ion Activity, Section 5
Statistical Tables, Section 1
Statistics, Section 1
Steroid Hormones, Section 7

Steroids, Hormonal, Section 7
Steroids, Properties of, Section 7
Stock Solutions, Standard, Section 11
Straight-Chain Alkanes, Names of, Section 7
Substances, Pure, Physical Properties of, Section 7
Substituent Constants, Hammett and Taft, Section 3
Substitutive Nomenclature, Characteristic Groups for, Section 7
Sucrose Solutions, Aqueous, Section 10
Suffixes for Specialist Nomenclature of Heterocyclic Systems, Section 7
Sulfuric Acid, Aqueous Tensions of Aqueous Solutions of, Section 10
Sulfuric Acid, Relative Humidities of Aqueous Solutions of, Section 10
Surface Areas and Volumes, Section 1
Surface Tension, Section 10
Surface Tension of Selected Inorganic Substances, Section 10
Surface Tension of Water, Section 10
Sustituent Groups on Alkyl Chemical Shifts, Effect of, Section 8
Symbols and Abbreviations, Mathematical, Section 2
Symbols, Standard Letter, Section 2
Synonyms and Mineral Names, Section 4
Symbols and Terminology, Physical and Chemical, Section 2
Synthetic and Natural Rubbers, Properties of, Section 7
Synthetic Rubbers, Section 7
Synthetic Rubbers, Formulas of, Section 7
Synthetic Waxes, Section 7

T
t, Values of, Section 1
Table of Nuclides, Section 3
Taft Substituent Constants, Section 3
Taft Equation, pK_a^0 and p Values for, Section 3
Taft Equations, Section 3
Target Elements, ß Filters for, Section 8
Temperature and Its Measurement, Section 11
Temperature Conversion Table, Section 2
Temperature Correction, Barometer, Section 2
Temperature Dependence of Equilibrium Constants in Aqueous Solutions, Section 5
Temperature, Molar Equivalent of One Liter of Gas, Section 2
Temperature, Potentials of Electrodes in Volts as a Function of, Section 5
Temperature, Specific Gravity of Air, at, Section 10
Temperatures and Pressure, Section 2

Temperatures, Barometric Pressures at, Boiling Points of Water,
    Section 10
Temperatures, Conductivity of Very Pure Water at, Section 6
Temperatures, Values of 2.3026RT/F at, Section 5
Terminology, Physical and Chemical, Section 2
Ternary Azeotropes Containing Water and Alcohols, Section 10
Theoretical plates needed to achieve desired degrees of purity, Section 5
Thermal Conductivity, Section 10
Thermal Conductivity of Various Gases and Vapors, Section 10
Thermal Conductivity of Various Liquids and Solutions, Section 10
Thermal Conductivity of Various Metals and Alloys, Section 10
Thermal Expansion and Compressibility, Section 10
Thermal Expansion, Cubical Coefficients of, Section 2
Thermocouples, Section 11
Thermocouples, Thermoelectric Valves at the Fixed Points for,
    Section 11
Thermodynamic Properties, Section 9
Thermodynamic Relations, Section 9
Thermoelectric Valves at the Fixed Points for Various Thermocouples,
    Section 11
Thermometer Scales, Conversion, Section 2
Thermometers, Correction for Emergent Stem of Liquid-in-Glass,
    Section 11
Thermoplastic and Thermosetting Materials, Formulas of, Section 7
Thermoplastic and Thermosetting Materials, Properties of, Section 7
Thermosetting and Thermoplastic Materials, Formulas of, Section 7
Thermosetting and Thermoplastic Materials, Properties of, Section 7
Titrations, Aqueous, Acid-Base, and Colorimetric Determinetion of pH,
    Indicators for, Section 5
Titrimetric (Volumetric) Analysis, Equations and Equivalents for,
    Section 5
Titrimetric (Volumetric) Factors, Section 5
Titrimetric (Volumetric) Solutions, Section 5
Tolerances for Analytical Weights, Section 2
Tolerances for Volumetric Burets and Pipets, Section 2
Transmittance-Absorbance Conversion Table, Section 2
Trigonometric Functions of an Angle, Section 1
Trivial Names of Heterocyclic Systems Suitable for Use in Fusion
    Names, Section 7
Trivial Names for Hetero. Systems that are *not* for Use in Fusion
    Names, Section 7
Type B Thermocouples: Part-30% Rh Alloy vs. Part-6% Rh Alloy,
    Section 11

Type E Thermocouples: Ni-Cr Alloy vs. Co-Ni Alloy, Section 11
Type J Thermocouples: Ni-Cr Alloy vs. Ni-Al Alloy, Section 11
Type K Thermocouples: Part-13% Rh Alloy vs. Part, Section 11
Type T Thermocouples: Cu vs. Cu-Ni Alloy Changes in Calibration,
    Section 11

U
U.S. Sieve Series, Section 11
Units, Prefixes for Naming Multiples and Submultiples of, Section 2

V
Values of Absorbance for Percent Absorption, Section 2
Values of t, Section 1
Values of 2.3026RT/F at Several Temperatures, Section 5
Van der Waals' Constants for Gases, Section 11
Vapor-Pressure Equations, Section 10
Vapor Pressure of Ice in Millimeters of Mercury, Section 10
Vapor Pressure of Liquid Ammonia, NH_3, Section 10
Vapor Pressure of Mercury, Section 10
Vapor Pressures of Various Organic Compounds, Section 10
Vapor Pressure of Water in Millimeters of Mercury, Section 10
Vapor Pressures, Section 10
Vapor Pressures of Various Inorganic Compounds, Section 10
Vapors and Gases, Thermal Conductivity of, Section 10
Various Temperatures and Pressure, Section 2
Viscosity, Section 10
Viscosity Conversion Table, Section 2
Viscosity of Selected Inorganic Substances, Section 10
Viscosity of Water, Section 10
Vitamins, Section 7
Voltaic Cells and Batteries, Section 6
Volts, Electrodes in, as a Function of Temp., Potentials of Section 5
Volts for Water-Org. Solvent Mixtures, Potentials of Ref. Electrodes in
    Section 5
Volume, Factors for Simplified Computation of, Section 2
Volumes, Surface, Section 1
Volumetric Burets and Pipets, Tolerances for, Section 2
Volumetric (Titrimetric) Analysis, Equations and Equivalents for,
    Section 5
Volumetric (Titrimetric) Factors, Section 5
Volumetric (Titrimetric) Solutions, Section 5

W

W La$_1$ Line and Ka$_1$ Lines, Mass Absorption Coefficients for, Section 8
Water and Alcohols, Ternary Azeotropes Containing, Sec .10
Water and Mercury, Density of, Section 10
Water, Binary Azeotropes Containing, Section 10
Water, Boiling Points of at Pressures, Section 10
Water, Boiling Points of, Barometric Pressures at Temperatures, Section 10
Water, Compressibility of, Section 10
Water Conductivity at Various Temperatures, Section 6
Water, Dielectric Constant of, Section 10
Water, Ionic Production Constant of, Section 5
Water, Mercury, and Aluminum Oxide, Heat Capacity Standards, Section 9
Water-Organic Solvent Mixtures, Potentials of Ref. Electrodes in Volts for, Section 5
Water, Refractive Index of, Section 10
Water, Solubilities of Inorganic Compounds at Various Temperatures, Section 10
Water, Solubilities of Metal Salts at Various Temperatures, Section 10
Water, Solubilities of Organic Acids at Various Temperatures, Section 10
Water, Solubility of Gases in, Section 10
Water, Surface Tension of, Section 10
Water Vapor in Saturated Air, Mass of, Section 10
Water, Vapor Pressure of, in Millimeters of Mercury, Section 10
Water, Viscosity of, Section 10
Wavelength/Wavenumber Conversion Table, Section 2
Wavelength of Absorption Edges in Angstroms, Section 8
Wavelengths of X-ray Emission Spectra in Angstroms, Section 8
Wavenumber/Wavelength Conversion Table, Section 2
Waxes, Constants of, Section 7
Waxes, Synthetic, Section 7
Weighings in Air to Weighings in Vacuo, Conversion of, Section 2
Weights, Tolerances for, Section 2
Wet and Dry Bulb Readings, Humidity and Dew Point from, Section 10

X

X-Ray Absorption Energies in KeV, Section 8
X-Ray Emission Energies in keV and Filters, Section 8
X-Ray Emission Spectra in Angstroms, Wavelengths of, Section 8
X-Ray Methods, Section 8
X-Ray Spectroscopy, Analyzing Crystals for, Section 8

B.23.
Literature Sources

Books

Bottle, R. T., Rowland, J. F. B. (eds.). *Information Sources in Chemistry*. London, New York: Bowker-Saur, 1993 (QD8.5.I47 1993).

Dorman, Phae H. *Chemical Industries: An Information Sourcebook*. Phoenix: Oryx, 1988 (Z5521.D65 1988).

Maizell, Robert Edward. *How to Find Chemical Information*. 2nd ed. New York: Wiley, 1987 (QD8.5.M34 1987).

Mellon, M. G. *Chemical Publications: Their Nature and Use. 5th ed.* New York: McGraw-Hill, 1982 (QD8.5.M44 1982).

Pickering, W. R. *Information Sources in Pharmaceuticals*. London, New York: Bowker-Saur, 1990.

Skolnik, H. *The Literature Matrix of Chemistry*. New York: Wiley-Interscience, 1982.

Wiggins, Gary. *Chemical Information Sources*. New York: McGraw-Hill, 1991 (QD8.5.W54 1991).

Wolman, Yecheskel. *Chemical Information: A Practical Guide to Utilization*. Chichester, New York: Wiley, 1988 (QD8.5.W64 1988).

Woodburn, H. M. *Using the Chemical Literature: A Practical Guide (Library and Information Science Series, Vol 11)*. New York: Marcel Dekker, Inc., 1974.

Articles

Hancock, J. E. H. An Introduction to the Literature of Organic Chemistry. *J Chem Ed* **45**; 193, 260, and 336 (1968).

Somerville, Arleen N. Information Sources for Organic Chemistry. *J Chem Ed* **68**; 553 and 842 (1991); **69**: 379 (1992).

Carr, Carol. Teaching and Using Chemical Information. *J Chem Ed* **70**; 719 (1993).

Internet Resources

ChemCenter; A Service of the American Chemical Society

http://www.ChemCenter.org

Chemical Information Sources from Indiana University: CIS-IU (Gary Wiggins)

http://www.indiana.edu/~cheminfo/

InfoSurf: Chemistry

> http://www.library.ucsb.edu/subj/chemistr.html

WebChemistry

> http://www.latrobe.edu.au/www/webchem/

World-Wide Web Virtual Library: Chemistry

> http://www.chem.ucla.edu/chempointers.html

B.24.

The Merck Index:

RS356.M47

- **compound/trade name and structure**
- **preparation and derivatives of compounds**
- **physical and pharmaceutical properties**
- **organic name reactions**
- **conversion and mathematical tables**

B.24.a.   OVERVIEW

A source of information about chemicals of medicinal importance. The Index contains structural diagrams, key references, use, and toxicity data.

EASY ACCESS

— Formula Index, located in the back of the book, provides monograph number.
— Cross Index of Names, located in the back of the book, provides monograph number.
— Consult the alphabetical list of tables in section B.24.b to confirm the presence of a desired table in the Index.

INDEXING

The monograph section lists chemicals, drugs, and biological substances contained in the index by monograph number. Monograph numbers are assigned sequentially to the monograph titles as they appear in alphabetical order by generic, trivial, or simple chemical name. Monograph numbers do not correspond between editions of the **Merck Index**. Items in the other indices are referenced to these numbers.

The **Chemical Abstracts Service Registry Numbers Index** provides the Chemical Abstracts Service Registry Numbers for the monograph title compounds and selected derivatives. An asterisk next to the compound name indicates that it is a derivative of a title compound. The index is composed of two tables; the first lists the compounds in alphabetical order with their corresponding CAS number, and the second is ordered numerically by CAS registry number.

The **Therapeutic Category and Biological Activity Index** lists compounds by their therapeutic category and biological activities, including mechanism of action. A master list of all categories and cross-references can be found on the first few pages of the index. Monographs are listed alphabetically by title under each heading. Selected derivatives and isomers of title compounds have been listed by generic or trivial name and are referenced to the appropriate monograph by number. These listings are cross-referenced to the appropriate therapeutic categories and any synonyms of preferred terms or closely related entries. Many therapeutic category headings have been subclassified according to structural features.

The cumulative **Formula Index** lists the molecular formula of chemical substances with their monograph numbers. Formulas are arranged by citing carbons in ascending order, then hydrogens, and finally the remaining elements in alphabetical order. If carbon is not present, the elements are arranged in alphabetical order. An asterisk next to the compound name indicates that it is a derivative of a title compound.

The **Cross Index of Names** lists the chemical titles, synonyms, and miscellaneous trade names in alphabetical order with their monograph numbers. Trade names are listed with an abbreviated form of the company of ownership. An asterisk preceding an entry signifies that the name does not appear in the monograph.

COMPUTER SEARCH

The Merck Index Online is available through Dialog Information Retrieval Service (Database [304]) or through STN International (FILE MRCK). The Basic Index (BI) contains single words from the applications, Merck Index Name, CA Index Name, synonyms, trade names, associated company names, drug codes, therapeutic codes, therapeutic veterinary codes, toxicity organisms, toxicity dose, toxicity route, toxicity text, notes and caution, references, property data text fields and the related fields for the derivatives, as well as molecular formulas and weights and CAS Registry Numbers. Example search fields and display codes are given below.

Search Field Name	Search Code	Search Example	Display Code
Chemical name	/CN	S DURACEF/CN	CN
Chemical derivative	/CN.DRV	S VASOTEC/CN.DRV	CN.DRV
MERCK index name	/MIN	S LINOPIRDINE/MIN	MIN
Organism-tox test	/ORGN	S MICE/ORGN	TOX
Therapeutic codes	/THER	S ANTIDEPRE- -SSANT/THER	THER
Toxicity test	/TOX	S RAT#/TOX	TOX
Veterinary therapeutic codes	/VTHER	S ANTICOA- GULANT/VTHER	VTHER

The default display format, DISPLAY L(*ist number*), in MRCK provides the Merck number (MNO), CAS registry no. (RN), Merck index name (MIN), CA index name (CN), synonyms (CN), molecular formula (MF) linear structural formula (LSF), weight composition (COMP), molecular weight (MW) and structural formula (STR). The DISPLAY ALL command also gives the references (RE), boiling point (BP), melting point (MP), refractive index (RI), optical rotation (OPR), specific gravity (SPGR), toxicity (TOX), UV and visible spectrum (UVS), other properties (OCPP), notes (NTE), application (APP), therapeutic codes (THER), veterinary therapeutic codes (VTHER), other sources (OS), and referenced patent (RPN) for the compound and derivatives. Any of these individual data fields can be accessed with the DISPLAY command.

(Copyright 1998 by the American Chemical Society and reprinted with permission.)

CD-ROM

The *Merck Index on CD-ROM* contains all the monographs in the printed edition as well as additional monographs compiled after publication, supplementary tables, and Organic Name Reactions. It can be searched by text, structure, or substructure. The CD-ROM provides the tools to draw and search for a chemical structure.

INTERNET

STN MRCK Database Summary Sheets can be accessed at:

> http://info.cas.org/ON LINE/DBSS/mrckss.html

Dialog Information Retrieval Service Blue Sheets can be accessed at:

> http://library.dialog.com/bluesheets

The Merck Index is published by the Merck Publishing Group of Merck & Co., Inc. Their web site is http://www.Merck.com.

ORGANIZATION

The compounds are arranged alphabetically by generic, trivial, or simple chemical name throughout the Index. Each entry is assigned a sequential monograph number to correspond to the monograph title. Monograph numbers do not correspond between editions of the **Merck Index**. Items in the other indices are referenced to by their monograph number. Trivial or chemical names are preferred because headings and trade names are used in limited instances.

EXAMPLE ENTRY

Title

Chemical Abstracts name
(boldface italic)

Monograph number

5460. Lovastatin. *[1S-[1α(R*),3α,7β,8β(2S*,4S*),8aβ]]-2-Methylbutanoic acid 1,2,3,7,8,8a-hexahydro-3,7-dimethyl-8-[2-(tetrahydro-4-hydroxy-6-oxo-2H-pyran-2-yl)ethyl]-1-naphthalenyl ester;* (1S,3R,7S,8S,8aR)-1,2,3,7,8,8a-hexahydro-3,7-dimethyl-8-[2-[(2R,4R)-tetrahydro-4-hydroxy-6-oxo-2H-pyran-2-yl]ethyl]-1-naphthalenyl (S)-2-methylbutyrate; 1,2,6,7,8,8a-hexahydro-β,δ-dihydroxy-2,6-dimethyl-8-(2-methyl-1-oxobutoxy)-1-naphthaleneheptanoic acid δ-lactone; 2β,6α-dimethyl-8α-(2-methyl-1-oxobutoxy)-mevinic acid lactone; mevinolin; 6α-methylcompactin; monacolin K; MK 803; Mevacor; Mevinacor; Mevlor. $C_{24}H_{36}O_5$; mol wt 404.55. C 71.25%, H 8.97%, O 19.78%. Fungal metabolite, potent inhibitor of HMG-CoA reductase, the rate controlling enzyme in cholesterol biosynthesis. Isoln from *Monascus ruber:* A. Endo, *J. Antibiot.* **32**, 852 (1979); from *Aspergillus terreus:* R. L. Monaghan *et al.,* U.S. pat. **4,231,938** (1980 to Merck & Co.). Structure and biochemical properties: A. W. Alberts *et al., Proc. Nat. Acad. Sci. USA* **77**, 3957 (1980). Synthesis: M. Hirama, M. Iwashita, *Tetrahedron Letters* **24**, 1811 (1983); D. L. J. Clive *et al., J. Am. Chem. Soc.* **110**, 6914 (1988). Biosynthesis: M. D. Greenspan, J.B. Yudkovitz, *J. Bacteriol.* **162**, 704 (1985); R. N. Moore *et al., J. Am. Chem. Soc.* **107**, 3694 (1985). HPLC determn in plasma and bile: R. J. Stubbs *et al., J. Chromatog.* **383**, 438 (1986). Stimulation of receptor-medicated clearance of low density lipoproteins: D. W. Bilheimer *et al., Proc. Nat. Acad. Sci. USA* **80**, 4124 (1983). Effects on lipoprotein metabolism: S. M. Grundy, G. L. Vega, *J. Lipid Res.* **26**, 1464 (1985). Multicenter clinical comparison with gemfibrozil, *q.v.:* M. J. Tikkanen *et al., Am. J. Cardiol.* **62**, 35J (1988). Review of syntheses: T. Rosen, C. H. Heathcock, *Tetrahedron* **42**, 4909-4951 (1986). Review of clinical experience: J. A. Tobert, *Circulation* **76**, 534-538 (1987); *idem, Am. J. Cardiol.* **62**, 28J-34J (1988)

Drug code number

Molecular formula

Molecular weight

Literature references

Alternate chemical names (lightface)

Trademarks (capitalized)

Percentage composition

Patent and chemical information

Biological, pharmacological, clinical information, etc.

Review articles

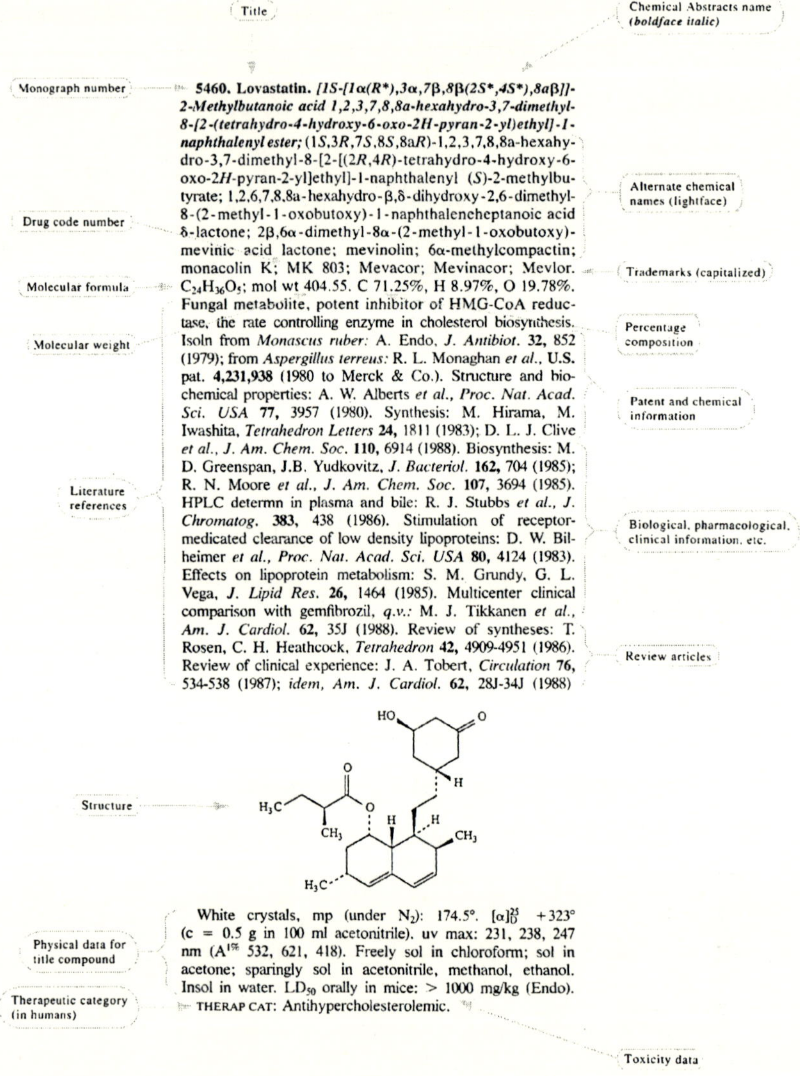

Structure

White crystals, mp (under N_2): 174.5°. $[\alpha]_D^{25}$ +323° (c = 0.5 g in 100 ml acetonitrile). uv max: 231, 238, 247 nm ($A^{1\%}$ 532, 621, 418). Freely sol in chloroform; sol in acetone; sparingly sol in acetonitrile, methanol, ethanol. Insol in water. LD_{50} orally in mice: > 1000 mg/kg (Endo). THERAP CAT: Antihypercholesterolemic.

Physical data for title compound

Therapeutic category (in humans)

Toxicity data

The Merck Index: An Encyclopedia of Chemicals, Drugs, and Biologicals, Twelfth Edition, Susan Budavari, Maryadele J. O'Neil, Ann Smith, Patricia E. Heckelman, Joanne F. Kinneary, Eds. (Merck & Co., Inc., Whitehouse Station, NJ, USA, 1996)

B.24.b.   ALPHABETICAL LIST OF TABLES IN THE MERCK INDEX

A
Alchemical symbols used in biology and botany
Air contaminants, maximum allowable concentrations

B
Boiling points of solvents
Buffers for calibration of pH meters

C
Cancer chemotherapy drug regimens
Company information
    Code letters for experimental substances
    Name and address directory
Concentration of acids and bases
Constant humidity solutions
Conversion factors, universal (units of weights and measures)
Cooling mixtures

G
Greek alphabet

I
Indicators
    Mixed indicators
    Volumetric work and pH determinations
International system of units (SI)
    SI basic and supplementary units
    SI derived units with special names
    SI prefixes
Isotonic solutions

L
Latin terms

M
Maximum allowable concentration of air contaminants
Molar volumes of laboratory chemicals

N
Numerical prefixes used in chemical names

P
Percentage solution tables (apothecary and metric)
pH values of standard solutions
Prescription notation

R
Radioactive isotopes
    Half-lives and radiations
    Use in medical therapy and diagnosis
Roman numerals
Russian alphabet

S
Saturated solutions
Solvents, boiling points
Specific gravity comparisons

T
Thermometric equivalents
    Conversion table (Celsius = Fahrenheit)
    Temperature scales and conversion equations

U
USAN-abbreviated terms for radicals

W
Weights of liquids (lb/gal; kg/l)
WHO-abbreviated terms for radicals

The Merck Index: An Encyclopedia of Chemicals, Drugs, and Biologicals, Twelfth
Edition, Susan Budavari, Maryadele J. O'Neil, Ann Smith, Patricia E. Heckelman,
Joanne F. Kinneary, Eds. (Merck & Co., Inc., Whitehouse Station, NJ, USA, 1996)

B.25.

Organic Reactions:

QD251.O6

- review articles
- example experimentals
- organic name reactions

The series is a collection of chapters describing all facets of organic reaction. This includes experimental conditions and parameters, scope and limitations, as well as structural and functional group effects. Each chapter includes summation tables of all known examples of the reaction and several detailed laboratory procedures.

EASY ACCESS

— The Chapter and Topic Index in the back of the latest volume is cumulative for all the volumes in the collection.

INDEXING

The **Author Index** links the names of authors and coauthors to the corresponding volume in which their chapter appears. The index in the latest volume is cumulative for the entire collection.

The **Chapter and Topic Index** lists the functional groups, reactions, and general chemical reagents with the corresponding volume number in bold type and chapter number where the entry appears.

There is also a listing of **Cumulative Chapter Titles by Volume** in the back of each volume.

ORGANIZATION

Organic Reactions is a collection of volumes beginning with Volume 1 in 1942. The volumes are published annually with cumulative author, chapter, and title indexes. A program of updating earlier reviews was initiated beginning with Volume 22 in 1975.

B.26.

Organic Synthesis:

QD262.O6

- **tested laboratory methods**
- **preparation of specific compounds**

Organic Synthesis is an annual compilation of checked laboratory methods that emphasize the preparation of model compounds with unique reactions or molecules of interest to organic researchers.

EASY ACCESS

— For an experimental procedure or compound, check the General Index in the *Collective Indices to Collective Volumes 1–8*.

— For a particular reaction, look in the *Reaction Guide to Collective Volumes 1–7 and Annual Volumes 65–68*. If it is not in the Reaction Guide, check the reaction index to *Collective Volume 8*.

INDEXING

The 12 different indices used in *Organic Synthesis, Collective Volumes I–VII, Cumulative Indices Volumes I–V*, and *Cumulative Indices Volumes I–VIII* are:

Index	Cumulative Indices	Collective Volumes
Apparatus	✓	
Author	✓	✓
Concordance		✓
Contents (by CA Name)	✓	
Contents (by Title Name)	✓	
Formula	✓	✓
General	✓	✓
Hazard & Waste Disposal	✓[a]	✓[b]
Tables	✓[a]	
Solvents & Reagents	✓	
Type of Compound	✓	✓
Type of Reaction	✓	✓

[a] Only in Collective Indices to Collective Volumes 1–8. [b] Collective Volume 6 and subsequent volumes.

The *Organic Synthesis, Reaction Guide* incorporates Collective Volumes 1–7 and Annual Volumes 65–68.

The **Author Index** links the names of authors who contributed procedures to the corresponding page where their experimental procedure appears. In the case of coauthors, each author is listed as a separate entry and the names of the checkers of the reaction conditions are not included in the index.

The **Apparatus Index** lists special or less common pieces of apparatus and equipment and the Collective Volume numbers in boldface and pages. Boldfaced page numbers are used to denote an illustration of the scientific apparatus.

The **Concordance Index** links the annual volume in boldfaced type and page number where each procedure first appeared with the page number for the revised procedure in the collective volume.

The **Cumulative Contents Index (according to Title Names of Preparation)** lists the preparations according to the annual volume main title names in boldfaced capital letters followed by the Collective Volume number (boldface) and page where the citation appears. The second line of each entry lists the Chemical Abstracts Index Name with the CAS Registry Number in square brackets.

The **Cumulative Contents Index (according to Chemical Abstracts Index Names)** lists the Chemical Abstracts Index Names, with the CAS Registry Number in square brackets. On the next line the Title Names are given in capital letters with the Collective Volume numbers (boldface) and pages where the preparation can be found.

The **Formula Index** links the molecular formula of chemical substances listed in the Type of Compound Index with the Title Name and page where the preparation can be found. Formulas are arranged by citing carbons in ascending order, then hydrogens, and finally the remaining elements in alphabetical order. If carbon is not present, the elements are arranged in alphabetical order. Organic acid salts are listed under the formulas of the parent acid.

The **General Index** includes references to reagents, chemicals, and catalysts used in the various preparations as well as to the intermediates and final products. The names are shown in capital letters with the Collective Volume and page numbers in boldfaced type, when the compound's preparation is described in detail. The name of a substance with a page number in boldfaced type is used for preparations that are adequate but not in full detail. If the compound is only mentioned in the entry, lightfaced type is used. The systematic Chemical Abstracts name appears below the title name of each substance.

The **Hazard and Waste Disposal Index** lists all chemicals and functional groups that present a hazard or danger in the laboratory with the volume page where safe handling and disposal instructions can be found.

The **Reaction Guide,** a separate volume, links a graphic representation of each Organic Synthesis preparation with the Collective or Annual Volume and page the experimental procedure is given. The reactions are indexed based on a system of 11 broad classes and related subclasses as described in the preface to the Guide.

The **Solvents and Reagents Index** links the names of common solvents and reagents with the Collective Volume numbers in boldface and pages where the entry can be found. The procedures describe the purification, preparation, or assay of various reagents and solvents.

The **Tables Index** links tables with examples of compounds or functional groups made by a described procedure with the volume in boldface and page where the citation can be found.

The **Type of Compound Index** is sectioned into the various functional groups and ring systems and then further divided into more descriptive headings such as aliphatic, aromatic, and heterocyclic aldehydes. Preparations are characterized by the functional group being introduced, and organic salts are included with the parent acid or base. The compound names with the Collective Volume in boldface and pages are listed under each divisional heading.

The **Type of Reaction** is an alphabetical listing of the various general types of reactions, functional groups, or named reactions. The sections are further divided into more descriptive headings such as addition to C=C or C=O. The preparation title with the Collective Volume in boldface and pages are listed under each divisional heading. An important reagent for the transformation may be given in square brackets before the Collective Volume number.

COMPUTER SEARCH

A searchable graphics database of **Organic Synthesis** called ORGSYN is available from MDL Information Systems, Inc., 14 Walsh Drive, Parsippany, NJ 07054; Phone: 800-401-4321.

INTERNET

A description of the **Organic Synthesis** database, ORGSYN, is available at the MDL Information Systems, Inc. web site:

http://www.mdli.com

ORGANIZATION

Organic Synthesis has published an Annual Volume every year since 1921. The procedures in the Annual Volumes are revised, improved and collected into the Collective Volumes at 5–to 10-year intervals.

Annual Volumes	Collective Volume	Other
1–9	I	
10–19	II	
20–29	III	
30–39	IV	
40–49	V	Collective Index to I–V
50–59	VI	
60–64	VII	Reaction Guide to I–VII
65–69	VIII	Collective Index to I–VIII

B.27.

Patai's Chemistry of Functional Groups Series:

• preparation and chemistry of functional groups

A series of advanced treatises covering theory, physical properties, synthesis, and reactivity of organic functional groups.

B.27.a. OVERVIEW

EASY ACCESS

— **Patai's 1992 Guide to the Chemistry of Functional Groups**, by Saul Patai (QD 215.2.P34 1992).
— Index of Titles (with call numbers), section B.27.b.

INDEXING

Each Functional Group Volume has an author and subject index located in the back of the book. If a Functional Group Volume is divided into "parts," the indices are located at the end of the last part in that volume. The **Author Index** lists the names of all the authors of references used in the individual chapters of the volume. It lists the author's name with the cited reference number in parentheses. Next, the page number where the citation appears is listed in normal type, followed by the page in italics where the reference is listed. Numerous citations by the same author are listed in order of ascending page number.

The **Subject Index** links references to functional groups, compounds, and physical properties with the page where the topic is described.

Patai's 1992 Guide to the Chemistry of Functional Groups, by Saul Patai (QD 215.2.P34 1992) indexes the entire series to 1992. It contains three major sections:

SECTION I lists the titles and tables of contents of all the main volumes and updates from 1964 to 1991.

SECTION II gives a chapter-by-chapter description of the subject matter in each of the volumes, including chapter authors and coverage period. A *complementary* list of chapters in other volumes of the series that deal with the similar subject matter relating to the same functional group and a *relevant* list of all chapters in the series that deal with the same type of material but relating to different functional groups is attached to each account.

SECTION III contains the author and chapter/subject indexes. The **Author Index** links the names of the chapter authors with the page where their chapter is described in sections I and II.

The **Chapter/Subject Index** connects the name or one or more key words characterizing each chapter about a functional group with the page in section II where the subject can be found.

ORGANIZATION

Each book in the series describes the general and theoretical aspects, reactions, preparation, spectroscopy, physical properties, and derivatives of a particular functional group.

B.27.b.   PATAI'S CHEMISTRY OF FUNCTIONAL GROUPS, INDEX OF TITLES

Functional Group	The Chemistry of	Call Number
A		
acetals	Supplement E: ethers, crown ethers, hydroxyl groups, and their sulfur analogues (1980)	QD305.E7 C48
acetylenes	carbon-carbon triple bond (1978)	QD305.H8 C43
	Supplement C: triple-bonded functional groups (1983)	QD305.H8 C44
	Supplement C2: triple-bonded functional groups (1994)	QD305.H8 C518
acid anhydride	Supplement B: acid derivatives (1979)	QD305.A2 C494
acyl halides	acyl halides (1972)	QD305.H15 P37
	Supplement B: acid derivatives (1979)	QD305.A2 C494
alcohols	hydroxyl group (1971)	QD305.A4 P25
	Supplement E: ethers, crown ethers, hydroxyl groups, and their sulfur analogues (1980)	QD305.E7 C48
	Supplement E: hydroxyl, ether and peroxide groups, Vol. 2 (1993)	QD305.E7 C484
aldehydes	carbonyl group (1966)	QD305.A6 P318
	carbonyl group, Vol. 2 (1970)	QD305.A6 C47
aldehydes, α-halo α-haloketones, α-haloaldehydes, and α-haloimines (1988)		QD305.K2 K55

Functional Group	The Chemistry of	Call Number
alkanes	Alkanes and Cycloalkanes (1992)	QD305.H6 C44
alkenes	alkenes (1964)	QD305.H7 P3
	alkenes, Vol. 2 (1970)	QD305.H7 P3
	Supplement A: double-bond functional groups (1977)	QD476.C53
	Supplement A: double-bonded functional groups, Vol. 2 (1989)	QD461.C42
	Supplement A3: double-bonded functional groups (1997)	QD476.C53
alkynes	carbon-carbon triple bond (1978)	QD305.H8 C43
	Supplement C: triple-bonded functional groups (1983)	QD305.H8 C44
	Supplement C2: triple-bonded functional groups (1994)	QD305.H8 C518
allenes	ketenes, allenes, and related compounds (1980)	QD305.K2 C46
amides	amides (1970)	QD305.A7 Z17
amidines	amidines and imidates (1975)	QD341.A7 C63
amino compounds	amino group (1968)	QD305.A8 P37
	Supplement F: amino, nitroso, and nitro compounds and their derivatives (1982)	QD305.A8 C425
	Suppplement F2. Part 2: amino, nitroso, nitro compounds and related groups (1996)	QD305.A8 C425
antimony compounds	organic arsenic, antimony, and bismuth compounds (1994)	QD412.A7 C53

arsenic compounds	organic arsenic, antimony, and bismuth compounds (1994)	QD412.A7 C53
azides	azido group (1971)	QD305.A9 P37
	Supplement D: halides, pseudo-halides, and azides (1983)	QD165.C48
	Supplement D2: halides, pseudo-halides, and azides (1995)	QD165.C48
azido	azido group (1971)	QD305.A9 P37
azo compounds	hydrazo, azo, and azoxy groups (1975)	QD341.A9 P36
azomethines	carbon-nitrogen double bond (1970)	QD305.I6 P3
	hydrazo, azo, and azoxy groups (1975)	QD341.A9 P36
azoxy compounds	hydrazo, azo, and azoxy groups (1975)	QD341.A9 P36

B

bismuth compounds	organic arsenic, antimony, and bismuth compounds (1994)	QD412.A7 C53

C

carbon-carbon triple bond	carbon-carbon triple bond (1978)	QD305.H8 C43
	Supplement C: triple-bonded functional groups (1983)	QD305.H8 C44
	Supplement C2: triple-bonded functional groups (1994)	QD305.H8 C518
carbon-halogen bond	carbon–halogen bond (1973)	QD305.H15 P25
carbon-nitrogen double bond	carbon-nitrogen double bond (1970)	QD305.I6 P3

Functional Group	The Chemistry of	Call Number
carbonyl compounds	carbonyl group (1966) carbonyl group, Vol. 2 (1970)	QD305.A6 P318 QD305.A6 C47
carboxylic acids	carboxylic acids and esters (1969) Supplement B: acid derivatives (1979) carboxylic acids, esters, and their derivatives (1991) Supplement B: acid derivatives, Parts 1 & 2 (1992)	QD305.A2 P323 QD305.A2 C494 QD305.A2 O44 QD305.A2 C54
carboxylic esters	carboxylic acids and esters (1969) Supplement B: acid derivatives (1979) carboxylic acids, esters, and their derivatives (1991)	QD305.A2 P323 QD305.A2 C494 QD305.A2 O44
crown ethers	ether linkage (1967) Supplement E: ethers, crown ethers, hydroxyl groups, and their sulfur analogues (1980) crown ethers and analogues (1989)	QD305.E7 P26 QD305.E7 C48 QD305.E7 C76
cyanates	cyanates and their thio derivatives (1977)	QD305.A2 C495
cyano compounds	cyano group (1970) cyano group (1970) carbon-carbon triple bond (1978) Supplement C: triple-bonded functional groups (1983) Supplement C2: triple-bonded functional groups (1994)	QD305.N7 R3 QD305.N7 R3 QD305.H8 C43 QD305.H8 C44 QD305.H8 C518
cycloalkanes	alkanes and cycloalkanes (1992)	QD305.H6 C44

cyclopropanes	cyclopropyl group (1987)	QD305.H9 C48
	cyclopropane derived reactive intermediates (1990)	QD305.H9 B63
	cyclopropyl group, Vol. 2 (1996)	QD305.H9 C48
cyclic sulphides	synthesis of sulphones, sulphoxides, and cyclic sulphides (1994)	QD305.S6 S96

D

diazo compounds	diazonium and diazo groups (1978)	QD305.A9 C54
diazonium compounds	diazonium and diazo groups (1978)	QD305.A9 C54
disulphides	sulphonium group (1981)	QD341.S8 C45
double-bonded groups	Supplement A: double-bonded functional groups (1977)	QD476.C53
	Supplement A: double-bonded functional groups, Vol. 2 (1989)	QD461.C42
	Supplement A3: double-bonded functional groups (1997)	QD476.C53

E

enamines	enamines (1994)	QD305.A8 R36
enols	enols (1990)	QD301.C45
enones	enones (1989)	QD305.A6 C46
epoxides	ether linkage (1967)	QD305.E7 P26
	Supplement E: ethers, crown ethers, hydroxyl groups, and their sulfur analogues (1980)	QD305.E7 C48
	Supplement E: hydroxyl, ether, and peroxide groups, Vol. 2	QD305.E7 C484
esters	carboxylic acids and esters (1969)	QD305.A2 P323
	Supplement B: acid derivatives (1979)	QD305.A2 C494
	carboxylic acids, esters, and their derivatives (1991)	QD305.A2 O44

Functional Group	The Chemistry of	Call Number
ethers	ether linkage (1967)	QD305.E7 P26
	Supplement E: ethers, crown ethers, hydroxyl groups, and their sulfur analogues (1980)	QD305.E7 C48
	Supplement E: hydroxyl, ether, and peroxide groups, Vol. 2	QD305.E7 C484
G		
germanium compounds	organic germanium, tin, and lead compounds (1995)	QD412.G5 C47
H		
halides	carbon-halogen bond (1973)	QD305.H15 P25
	Supplement D: halides, pseudo-halides, and azides (1983)	QD165.C48
	Supplement D2: halides, pseudo-halides, and azides (1995)	QD165.C48
hydrazo compounds	hydrazo, azo, and azoxy groups (1975)	QD341.A9 P36
hydroxy compounds	hydroxyl group (1971)	QD305.A4 P25
	Supplement E: ethers, crown ethers, hydroxyl groups, and their sulfur analogues (1980)	QD305.E7 C48
	Supplement E: hydroxyl, ether, and peroxide groups. Vol 2	QD305.E7 C484
I		
imidates	amidines and imidates (1975)	QD341.A7 C63
imidic acid derivatives	amidines and imidates (1975)	QD341.A7 C63

imines, α-halo	α-haloketones, a-haloaldehydes, and α-haloimines (1988)	QD305.K2 K55
isocyanates	cyanates and their thio derivatives (1977)	QD305.A2 C495
isocyano compounds	cyano group (1970) Supplement C: triple-bonded functional groups (1983) Supplement C2: triple-bonded functional groups (1994)	QD305.N7 R3 QD305.H8 C44 QD305.H8 C518
isothiocyandtes	cyanates and their thio derivatives (1977)	QD305.A2 C495

K

ketenes	ketenes, allenes, and related compounds (1980)	QD305.K2 C46
ketones	carbonyl group (1966) carbonyl group, Vol. 2 (1970)	QD305.A6 P318 QD305.A6 C47
ketones, α-halo	α-haloketones, α-haloaldehydes, and α-haloimines (1988)	QD305.K2 K55

L

lactams	Supplement B: acid derivatives (1979) synthesis of lactones and lactams (1993)	QD305.A2 C494 QD305.A2 O33
lactones	Supplement B: acid derivatives (1979) synthesis of lactones and lactams (1993)	QD305.A2 C494 QD305.A2 O33
lead compounds	organic germanium, tin, and lead compounds (1995)	QD412.G5 C47

M

| metal-carbon bond | **see** organometallic compounds | |

N

| nitones | nitrones, nitronates, and nitroxides (1989) | QD305.N8 B83 |

Functional Group	The Chemistry of	Call Number
nitro compounds	nitro and nitroso groups, Part 1 (1969)	QD305.N8 F36
	nitro and nitroso groups, Part 2 (1970)	QD305.N8 F47
	Supplement F: amino, nitroso, and nitro compounds and their derivatives (1982)	QD305.A8 C425
	Suppplement F2. Part 2: amino, nitroso, nitro compounds, and related groups (1996)	QD305.A8 C425
nitronates	nitrones, nitronates, and nitroxides (1989)	QD305.N8 B83
nitroso compounds	nitro and nitroso groups, Part 1 (1969)	QD305.N8 F36
	nitro and nitroso groups. Part 2 (1970)	QD305.N8 F47
	Supplement F: amino, nitroso and nitro compounds and their derivatives (1982)	QD305.A8 C425
	Suppplement F2. Part 2: amino, nitroso, nitro, compounds, and their related groups (1996)	QD305.A8 C425
nitroxides	nitrones, nitronates, and nitroxides (1989)	QD305.N8 B83
O		
olefins	alkenes (1964)	QD305.H7 P3
	alkenes, Vol. 2 (1970)	QD305.H7 P3
	Supplement A: double-bond functional groups, (1977)	QD476.C53
	Supplement A: double-bonded functional groups, Vol. 2 (1989)	QD461.C42
	Supplement A3: double-bonded functional groups (1997)	QD476.C53

organic halides	carbon-halogen bond (1973)	QD305.H15 P25
	Supplement D: halides, pseudo-halides, and azides (1983)	QD165.C48
	Supplement D2: halides, pseudo-halides, and azides (1995)	QD165.G5 C47
organic Si compounds	organic silicon compounds (1989)	QD412.S6 P37
organic Se and Te compounds	organic selenium and tellurium compounds, Vol. 1 (1986)	QD412.S5 C53
	organic selenium and tellurium compounds, Vol. 2 (1987)	QD412.S5 C53
organometallic compounds	metal-carbon bond, Vol. 1 (1982)	QD410.C43
	metal-carbon bond, Vol. 2 (1985)	QD410.C43
	metal-carbon bond, Vol. 3 (1989)	QD410.C43
	metal-carbon bond, Vol. 4 (1987)	QD410.C43
	metal-carbon bond, Vol. 5 (1989)	QD410.C43
	organophosphorus compounds, Vol. 1 (1990)	QD412.P1 C444
	organophosphorus compounds, Vol. 2 (1992)	QD412.P1 C444
	organophosphorus compounds, Vol. 3 (1994)	QD412.P1 C444
	organophosphorus compounds, Vol. 4 (1996)	QD412.P1 C444
organoselenium compounds	organic selenium and tellurium compounds, Vol. 1 (1986)	QD412.S5 C53
	organic selenium and tellurium compounds, Vol. 2 (1987)	QD412.S5 C53
organosilicon compounds	organic silicon compounds (1989)	QD412.S6 P37

P

| peroxides | peroxides (1983) | QD181.O1 C46 |
| | Supplement E: hydroxyl, ether, and peroxide groups, Vol. 2 | QD305.E7 C484 |

Functional Group	The Chemistry of	Call Number
phosphines	organophosphorus compounds, Vol. 1 (1990)	QD412.P1 C444
phosphine oxides	organophosphorus compounds, Vol. 2 (1992)	QD412.P1 C444
phosphonium salts	organophosphorus compounds, Vol. 3 (1994)	QD412.P1 C444
phosphoranes	organophosphorus compounds, Vol. 3 (1994)	QD412.P1 C444
phosphorus acids	organophosphorus compounds, Vol. 4 (1996)	QD412.P1 C444
pseudo-halides	Supplement D: halides, pseudo-halides, and azides (1983) Supplement D2: halides, pseudo-halides, and azides (1995)	QD165.C48 QD165.C48

Q

Functional Group	The Chemistry of	Call Number
quinones	quinonoid compounds (1974) quinonoid compounds, Vol. 2 (1988)	QD341.K2 P37 QD341.Q4 C47

S

Functional Group	The Chemistry of	Call Number
selenides	organophosphorus compounds, Vol. 2 (1992)	QD412.P1 C444
selenium compounds	organic selenium and tellurium compounds, Vol. 1 (1986) organic selenium and tellurium compounds, Vol. 2 (1987)	QD412.S5 C53 QD412.S5 C53
silicon compounds	organic silicon compounds (1989) silicon-heteroatom bond: updates (1991)	QD412.S6 P37 QD181.S6 S443

sulfur compounds	Supplement S: sulphur-containing functional groups (1993)	QD305.S3 C48
sulphenic acids and derivatives	sulphenic acids and their derivatives (1990)	QD305.A2 C485
sulphides	synthesis of sulphones, sulphoxides, and cyclic sulphides (1994)	QD305.S6 S96
	organophosphorus compounds, Vol. 2 (1992)	QD412.P1 C444
sulphinic acids and derivatives	sulphinic acids, esters, and their derivatives (1990)	QD341.A2 C4175
sulphones	sulphones and sulphoxides (1988)	QD305.S6 C46
	synthesis of sulphones, sulphoxides, and cyclic sulphides (1994)	QD305.S6 S96
sulphonic acid and derivatives	sulphonic acids and their derivatives (1990)	QD305.A2 C485
	sulphonic acids, esters, and their derivatives (1991)	QD341.A2 C4176
sulphonium compounds	sulphonium group (1981)	QD341.S8 C45
sulphoxides	sulphones and sulphoxides (1988)	QD305.S6 C46
	synthesis of sulphones, sulphoxides, and cyclic sulphides (1994)	QD305.S6 S96

Functional Group	The Chemistry of	Call Number
T		
tellurides	organophosphorus compounds, Vol. 2 (1992)	QD412.P1 C444
tellurium compounds	organic selenium and tellurium compounds, Vol. 1 (1986)	QD412.S5 C53
thiocyanates	cyanates and their thio derivatives (1977)	QD305.A2 C495
thiols	thiol group (1974)	QD305.T45 P37
	Supplement E: ethers, crown ethers, hydroxyl groups, and their sulfur analogs (1980)	QD305.E7 C48
tin compounds	organic germanium, tin, and lead compounds (1995)	QD412.G5 C47
Y		
ylides	organophosphorus compounds, Vol. 3 (1994)	QD412.P1 C444

B.28.

Patents:

Books

Manual of Patent Examining Procedure. U.S. Department of Commerce, Patent and Trademark Office, Washington, D.C.; revised periodically.

The Derwent Guide to Patents. London: Derwent Publication Ltd., 1986 (T210.D47 1986).

Amernick, Burton A. *Patent Law for the Nonlawyer: A Guide for the Engineer, Technologist, and Manager.* New York: Chapman & Hall, 1991 (KF3114.8.E54 A44 1991b).

Auger, C. P., ed. *Information Sources in Patents.* London, New York: Bowker-Saur, 1992 (T210.I53 1992).

Carr, Fred K. *Essentials of the Patent.* Chapel Hill, N. C.: Patent Information, Inc., 1981 (KF3094.C37 1981).

Carr, Fred K. *Searching Patent Documents; for Patentability and Information.* Chapel Hill, N. C.: Patent Information, Inc., 1982 (T210.C37 1982).

Epstein, M. A. *Modern Intellectual Property*, 3rd ed. Englewood Cliffs, N. J.: Aspen Law and Business, 1995.

Fenner, Terrence W., Everett James L. *Inventor's Handbook.* New York: Chemical Publishing Co., 1969 (T212.F336).

Gordon, Thomas T., Cookfair, Arthur S. *Patent Fundamentals for Scientists and Engineers.* Boca Raton: CRC Lewis, 1995.

Grubb, Philip W. *Patents in Chemistry and Biotechnology.* 2nd ed. Oxford: Clarendon Press, 1986 (T211.G78 1986).

Grayson, Martin. *Information Retrieval in Chemistry and Chemical Patent Law.* New York: Wiley, 1983 (Z699.5 C5 I53 1983).

Knight, H. J. *Patent Strategy for Researchers and Research Managers*: New York, Wiley, 1996.

Maynard, John, T., Peters Howard, M. *Understanding Chemical Patents: A Guide for the Inventor*, 2nd ed. Washington, D. C.: American Chemical Society, 1991 (T211.M39 1991).

Reverdin, A., Schlaepfer, F., eds. *Katzarov's Manual on Industrial Property All Over the World.* Geneva: Katzarov SA, 9th ed., 1981; 11th rev., 1991.

Rimmer, B. M. *International Guide to Offlcial Industrial Property Publications*, 3rd ed. Revised and updated by S. Van Dulken. London: British Library, 1992.

Rosenberg, P. D. *Patent Law Basics.* Deerfield, Ill.: Clark Boardman Callaghan, 1992; revised annually.

Simmons, E. S., in C. J. Armstrong and J. A. Large, eds., *Manual of Online Search Strategies*, 2nd ed. Aldershot, U. K.: Ashgate, 1992, pp. 51–127.

Van Dulken, S. ed. *Introduction to Patents Information*, 2nd ed. London: British Library, 1992.

Warr, W., Suhr, C. *Chemical Information Management.* New York: VCH Publishers, 1992.

Wherry, Timothy Lee. *Patent Searching for Librarians and Inventors.* Chicago: American Library Association, 1995 (T210.W44 1995).

WIPO. Glossary of Terms Concerning Industrial Property Information and Documentation. *World Pat Inf 16*: 21–39 (1993).

Articles

"The Paradox of Patentability Searching." Edlyn S. Simmons. *Journal of Chemical Information and Computer Science*, 1985; 25, p. 379–386.

"Information and Documentation." Wendy A. Warr and Claus Suhr. *Ullmann's Encyclopedia of Industrial Chemistry*, 1990, B1–12.

"The Idiot's Guide to Patent Resources on the Internet." Nancy Lambert. *Searcher* 1995, 3, 34.

"More Patents—Lots More—on the Internet." Nancy Lambert. *Searcher* 1995, 3, 24.

"Patents, Literature." E. S. Simmons and S. M. Kaback. *Kirk-Othmer Encyclopedia of Chemical Technology*, 4th ed., 1996, 18, 102.

Government Publications

Official Gazette of the United States Patent and Trademark Office, U.S. Department of Commerce, Patent and Trademark Office. Published weekly. (Gov't Pubs U.S.C21.5)

Official Gazette of the United States Patent Office. Monthly Catalog. (Gov't Pubs U.S.C51.9)

Government Report Announcement and Index (Gov't Pubs U.S.C51.9)

Aluri, Rao; Robinson, Judith Schiek. A Guide to U.S. Government Scientific and Technical Resources. Littleton, CO: Libraries Unlimited, 1983. (Q224.3.U6A43 1983)

Schlag, Gretchen A. How to Get It: A Guide to Defense-Related Information Resources.Alexandria, VA: Defense Technical Information Center, 1989 (D7.15/2:89/1).

Schwartz, Julia. Easy Access to Information in United States Government Documents. Chicago, London: American Library Association,
1986 (Z1223.Z7S33 1986).

Databases

STN International offers the following patent databases:

CAS Patent Markush File 1988–Present	MARPAT
East German Patents from 1982–Present	PATDD
German Patent Database from 1968–Present	PATDPA
German Patents and Utility Models from 1968–Present	PATOSDE
IFI Patent Database from 1950–Present	IFIPAT
Japanese Patents from 1976–Present	JAPIO
Patents of European Patent Office from 1978–Present	PATOSEP
PCT Patents from 1983–Present	PATOSWO
U.S. Patents from 1971–Present	USPATFULL
World Patents Index from 1963–Present	WIPDS, WPINDEX

Dialog Information Retrieval Service offers the following patent
databases:
Claims/Citation [220, 221, 222]
Claims/Compound Registry [242]
Claims/Reassignment & Reexamination [123]
Claims Reference [124]
Claims/Uniterm [341, 223, 224, 225]
Claims/U.S. Patents [340, 23, 24, 25, 125]
Chinese Patent Abstracts in English [344]
Derwent Patent's Citation Index [342]
Derwent World Patents Index [350,351]
European Patents Full Text [348]
IMSWorld Patents International [447,947]
Japanese Patents [347]
U.S. Patents Full Text [654, 653, 652, PATFULL]

For the STN databases, the Basic Index (BI) contains single words
from the title (TI), abstract (AB), claims (CLM), detailed description (DETD), summary (SUMM), and drawing description (DRWD).
Example search fields and display codes are given below.

Search Field Name	Search Code	Search Example	Display Code
Chemical name	/CN	S DURACEF/CN	CN
Chemical derivative	/CN.DRV	S VASOTEC/CN.DRV	CN.DRV
MERCK index name	/MIN	S LINOPIRDINE/MIN	MIN
Organism-tox test	/ORGN	S MICE/ORGN	TOX
Therapeutic codes	/THER	S ANTIDEPRESSENT/THER	THER
Toxicity test	/TOX	S RAT#/TOX	TOX
Veterinary therapeutic codes	/VTHER	S ANTICOAGULANT/VTHER	VTHER

The default display format, DISPLAY L(*ist number*), provides the accession number (AN), title (TI), inventor (IN), patent assignee (PA), patent information (PI), application information (AI), related application information (RLI), document type (DT), line count (LN.CNT), Issue National Patent Classification Code (INCL), Current National Patent Classification Code (NCL), International Patent Classifications (IC), and examiner's field of search (EXF). The DISPLAY ALL command also gives the inventor address (INA), patent assignee address (PAA), patent assignee type (PAT), term of patent (PTERM), disclaimer date (DCD), priority information (PRAI), reference patent information (REP), reference non-patent information (REN), examiner name (EXNAM), legal representative (LREP), number of claims (CLMN), examplary claim number (ECL), number of pages for each PAGE format (GI), abstract (AB), government interest (GOVI), Parent Case Data (PARN), summary of the invention (SUMM), drawing description (DRWD), detailed description (DETD), patent claim text (CLM), Issue Main National Patent Classification Code (INCLM), Issue Secondary National Patent Classification Code (INCLS), Current Main National Patent Classification Code (NCLM), Current Secondary National Patent Classification Code (NCLS), IPC, Main (ICM), IPC, Secondary (ICS), and art unit (ARTU). Any of these individual data fields can be accessed with the DISPLAY command.

To find all equivalent patents to a CA patent record:
1. Find CAS abstract number of patent of interest
2. FILE CA
3. S *(CAS abstract number)*/AN
4. SEL PN (answer is a range (E) of numbers
5. FILE WPID
6. S E1
7. D PI

INTERNET RESOURCES

The Patent Information Users Group (PIUG) has a web site at:
http://piug.org. They list patent and IP offices, databases and search
servicers, patent and legal information, vendor sites, and frequently
asked questions on patents and legal issues. The European Patent Office
(EPO) has an extensive list of links to patent-related web sites at:
http://www.wpo.co.at/online/index.htm.

Country	Patent Office/Database
Argentina	www.cab.cnea.edu.ar/di
Australia	www.ping.at/patent
	ipaustralia.gov.au
Belgium	www.european-patent- office.org/patlib/country/belgium
Brazil	www.bdt.org.br/bdt/inpi
Canada	info.ic.gc.ca/ic-data/marketplace/cipo
	strategis.ic.gc.ca/sc_innov/patent/engdoc/ cover.html
China, People's Republic of	www.cpo.cn.net
Czechoslovakia	www.upv.cz
Denmark	www.dkpto.dk
European Patent Organization	www.epo.co.at/epo europa.eu.int/agencies/ohim/ohim.htm
Finland	www.prh.fi
France	www.inpi.fr
Georgia	www.global-erty.net/saqpatenti
Germany	www.deutsches-Patentamt.de
Greece	patlib/country/greece/index.htm

Hong-Kong	pluto.houston.com.hk/hkgipd/treaties.html
Hungary	www.hpo.hu
Indonesia	www.patent.go.id
Italy	www.european-patent-office.org/it/homepage
(FILPAT)	www.fildata.it
Japan	www.jpo-miti.go.jp
	www2.jpo-miti.go.jp
(JAPIO)	www.japio.or.jp/japio.htm
(ISTA)	www.intlscience.com
Korea	www.ik.co.kr/kopatent
Lithuania	www.is.lt/vpb/engl
Luxembourg	www.etat.lu/EC
Malaysia	kpdnhq.gov.my
Moldova	www.cri.md
Monaco	www.european-patent-office.org/patlib/country/monaco/index.htm
New Zealand	www.govt.nz/ps/min/com/patentOffice
Peru	ekeko.rcp.net.pe/INDECOPY
Poland	www.ibspan.waw.pl
Portugal	www.inpi.pt
Romania	www.osim.ro
Slovak Republic	www.indprop.gov.sk
Slovenia	www.sipo.mzt.si
Spain	www.eunet.es/InterStand/patentes
Sweden	www.prv.se/prveng/front.htm
Switzerland	www.ige.ch
United Kingdom	www.netwales.co.uk/ptoffice
	portico.bl.uk/services/sris/history.html
United States of America	www.uspto.gov
(IBM site)	www.ibm.com/patents
(MicroPatent)	www.micropat.com
(QPAT)	www.qpat.com
(CNIDR)	patents.cnidr.org
World Intellectual Property Orgn	www.wipo.org
	itl.irv.uit.no/trade_law/documents/i_p/wipo/art/wipo.html
(PCT)	pctgazette.wipo.int

B.29
Purchasing Chemicals

Books

Chem Sources—U.S.A. Clemson, S.C.: Directories Publishing Company, Inc.

Directory of Chemical Producers. Menlo Park, Calif: SRI International.

Specialty Chemicals Source Book. Endicott, N.Y.: Synapse Information Resources.

CD-ROM

The *Chem Sources—CD-ROM for Windows* contains all the chemicals available from Chem Sources—USA, International, and Europe editions. It can be searched by text, CAS registry number, or empirical formula.

Data Bases

Chemical Sources database is available through STN International by typing FILE CSCHEM. The Basic Index contains single words from the chemical name (CN), company name (CO), and trade name classifications (TNC), as well as company codes (COC) and CAS Registry Numbers (RN). Example search fields and display codes are given below.

Search Field Name	Search Code	Search Example	Display Code
Accession Number	/AN	S 102:123456	
CAS Registry Number	/RN	S 89123-39-7/RN	RN
Chemical Name	/CN	S BENZENE/CN	CN
Company Name	/CO	S ALDRICH/CO	CO
Trade Name Classifications	/TNC	S ADDITIVES/TNC	TNC

The default display format, DISPLAY L(*ist number*), in CSCHEM provides information on the accession number (AN), CAS Registry Numbers (RN), chemical name (CN), company codes (COC), and structure diagram (STR). The DISPLAY ALL command also gives the company name (CO) and trade name classifications (TNC). Any of these individual data fields can be accessed with the DISPLAY command.

INTERNET RESOURCES

Chemical Sources Online

http://www.chemsources.com

ChemConnect Chemicals Exchange

http://www.chemconnect.com

Chemcyclopedia

http://pubs.acs.org/chemcy/cen/chemcy.html

ChemExper

http://www.chemexper.be

ChemFinder

http://chemfinder.camsoft.com

Chemsite

http://www.chemsite.com

World Wide Web Chemicals

http://www.chem.com

B.30.

Reagents for Organic Synthesis (Fieser's)

QD262.F46

- **preparation and purification of compounds**
- **reactions of organic reagents**

EASY ACCESS

— The *Aldrich Catalog Handbook of Fine Chemicals* gives the Fieser reference for the chemicals they list.
— To find the reagent for a particular transformation, consult the appropriate index in the Cumulative Index, then all volumes after Volume 12.

INDEXING

The **Cumulative Index** covers Volumes 1–12. All subsequent volumes contain an Author Index and Subject Index covering only the particular treatise.

The **Author Index** links the names of author to the corresponding volume (bold type) and page where their reference appears. In the case of coauthors, each author is listed as a separate entry.

The **Synthesis Index** is divided into general categories by FUNC-TIONAL GROUPS (bold type), which are further classified by trans-formations based on the use of the group as either the product or the starting material. The principal reagents are given with the volume (bold type) and page where the entry can be found. The index also classifies the reagents used to prepare and react various heterocycles, difunctional molecules, and chiral compounds. Halo refers to all halogens with the particular entry differentiating the elements.

The **Type of Compound Index** relates structurally similar reagents to their corresponding chemical name. The CATEGORIES (bold type) include chiral reagents, catalysts, and named reactions and reagents. Those organometallic compounds whose names are complex and difficult to locate can be found in the METAL CONTAINING COMPOUNDS section.

The **Type of Reaction Index** lists GENERAL REACTION CATE-GORIES further divided into functional group transformations and named reactions. The principal reagents for each reaction are given with the volume (bold type) and page where the entry can be found.

The **Reagents Index (Subject Index)** lists the chemical reagents with the corresponding volume (bold type) and page where the entry appears.

ORGANIZATION

Fieser and Fieser' Reagents for Organic Synthesis is a 17-volume set. The reagents are listed alphabetically with the structural formula, physical constants, methods of preparation, purification, and applications for each entry. Each volume has its own an author index and reagent index.

B.31.

Rodd's Chemistry of Carbon Compounds

QD251.R7

- **properties and reactivity of *classes* of compounds**
- **preparation and properties of specific compounds**

B.31.a.   OVERVIEW

EASY ACCESS

— Refer to Section B.31.b Index of the Table of Contents for the desired volume, then consult the volume's index for the topic of interest.
— Cumulative Indexes are IG, IIE, and IIIH.
— Consult the detailed Table of Contents in the front of each volume for the class of compounds under investigation.

INDEXING

Each part has its own index covering the functional groups, reactions, and compounds mentioned in the text. They are arranged in alphabetical order. The Cumulative Indexes are:

IG: Aliphatic Compounds;
IIE: Alicyclic Compounds;
IIIH: Aromatic Compounds.

The **Index** links the chemical compound, reaction, or technique, with the volume and page where the information is cited. Chemical compounds have been indexed alphabetically under the names used by the authors and numbers that appear in boldface indicate that the compound is the title of the entry.

INTERNET

The volume titles are listed at:

http://www.elsevier.com/catalogue/SAA/series/RCC/Menu.html

ORGANIZATION

The treatise consists of five volumes with corresponding supplements. The first four volumes cover the preparation, properties, and principal reactions of carbon compounds, with extensive but selective bibliographies including reviews and surveys. Those topics covered in each volume are outlined on its spine.

Volume I: Aliphatic Compounds
A. General Introduction; Hydrocarbons; Halogenated Derivatives
B. Monohydric Alcohols, Their Ethers; Sulphur Analogues; Nitrogen Derivatives; Organometallic Compounds
C. Monocarbonyl Derivatives of Aliphatic Hydrocarbons; Their Analogues and Derivatives
D. Dihydric Alcohols, Their Oxidation Products and Derivatives
E. Trihydric Alcohols, Their Oxidation Products and Derivatives
F. Penta- and Higher-hydric Alcohols, Their Oxidation Products and Derivatives; Saccharides
G. Tetrahydric alcohols, their **analogues**, derivatives, and oxidation products; *Cumulative Index Volume I, Parts A–G*

Volume II: Alicyclic Compounds
A. Monocarbocyclic compounds C_3–C_5
B. Six- and higher-membered monocyclic compounds
C. Polycarbocyclic compounds, excluding steroids
D. Steroids
E. Steroids (continued); *Cumulative Index Volume II, Parts A–E.*

Volume III: Aromatic Compounds
A. General Introduction. Mononuclear hydrocarbons and their halogen derivatives, and derivatives with nuclear substitutents attached through nonmetallic elements from group VI of the Periodic Table
B. Benzoquinones and related compounds: Derivatives of mononuclear benzenoid hydrocarbons with nuclear substitutents attached through an element other than the non-metals in Groups VI and VII of the Periodic Table
C. Nuclear-substituted benzenoid hydrocarbons with more than one nitrogen atom in a substituent group
D. Derivatives of mononuclear benzenoid hydrocarbons with substituted aliphatic side-chains
E. Monobenzenoid hydrocarbon derivatives with functional groups in separate side-chains; dihydric and polyhydric aralkanols and their oxidation products; monobenzenoid hydrocarbons with unsaturated side-chains and their derivatives

F. Polybenzenoid hydrocarbons and their derivatives; hydrocarbon ring assemblies, polyphenyl-substituted aliphatic hydrocarbons and their derivatives
G. Monocarboxylic acids of the benzene series: C7–C13-carbocyclic compounds with fused-ring systems and their derivatives
H. Polycarbocyclic compounds with more than 13 atoms in the fused ring system. *Cumulative Index to Volume III A-H.*

Volume IV: Heterocyclic Compounds
A. Three-, four- and five-membered heterocyclic compounds with a Single Hetero-Atom in the ring
B. Five-Membered heterocyclic compounds with a Single Hetero-Atom in the Ring: Alkaloids, Dyes, and Pigments
C. Five-Membered Heterocyclic Compounds with Two Hetero-Atoms in the Ring from Groups V and/or VI of the Periodic Table
D. Five-Membered Heterocyclic Compounds with More than Two Hetero-Atoms in the Ring
E. Six-Membered Mono-Heterocyclic Compounds Containing Oxygen, Sulphur, Selenium and Tellurium, Silicon, Germanium, Tin, Lead, or Iodine as the Hetero-Atom
F. Six-Membered Heterocyclic Compounds with a Single Nitrogen Atom in the Ring; Pyridine, Polymethylenepyridines, Quinoline, Isoquinoline, and their Derivatives
G. Six-Membered Mono-Heterocycles Containing N, P, As, Sb, or Bi: Alkaloids with a Six-Membered Heterocyclic Ring
H. Alkaloids and Heterocycles Containing One Nitrogen Atom in the Ring (Contd.). Six-Membered Heterocycles with Two Hetero-Atoms in the ring
IJ . Six-Membered Heterocyclic Compounds with Two Hetero-Atoms from Group V of the Periodic Table, and with Three or more Hetero-Atoms
K. Six-Membered Heterocyclic Compounds with Two or more Hetero-Atoms, One or more of which are from Groups II, III, IV, V, or VII of the Periodic Table. Heterocyclic Compounds with Seven or More Atoms in the Ring
L. Fused-Ring Heterocyclic with Three or More Nitrogen Atoms. Biosynthesis of Alkaloids and Nitrogenous Microbial Metabolites.

B.31.b.　INDEX TO THE TABLE OF CONTENTS OF RODD'S CHEMISTRY OF CARBON COMPOUNDS

A
acenaphthene group, **IIIH**
acepleiadene, **IIIH**
acepleiadylene, **IIIH**
acetals, **IE**
acetic acid, hydroxy diphenyl, **IIIF**
acetylene oxides, **IVA**
acetylene, phenyl, **IIIE**
acetylenes, **IE** supplement
acids, **IC**, cyclic **IIA, IIB**
acids, aldehydic, **IIIE**
acids, amino, **ID**
acids, arene, **IIIE**
acids, di-, **ID**
acids, hydroxy, **ID**
acids, keto, **ID**
acids, nitro, **ID**
aconitine alkaloids, **IVG**
acridine alkaloids, **IVG**
acridine, **IVG**
acridone alkaloids, **IVG, IVL**
acridones, **IVL**
acyl arenes, **IIIE**
acyl azides, **IC**
acyl halides, **IC**
adamantane, **IIC**
agar, **IFG** supplement
alantolactone, **IIC**
alcohols, **IB**, cyclic **IIA, IIB**
alcohols, aralkyl, **IIID**
alcohols, aromatic, **IIIE**
alcohols, tetrahydric, **IG**
alcohols, trihydric, **IE**
aldehydes, **IC**, cyclic **IIA, IIB**
aldehydes, aromatic, **IIID**
aldehydes, arene di-, **IIIE**
aldehydes, di-, **ID**
aldehydes, di-, dihydroxy, **IG**
aldehydes, di-, hydroxy, **IE**
aldehydes, di-, keto, **IE**

aldehydes, dihydroxy, **IE**
aldehydes, hydroxyl, **ID**
aldehydes, ketohydroxy, **IE**
aldehydes, trihydoxy, **IG**
aldehydes, unsaturated, aromatic, **IIIE**
aldehydic acids, **IIIE**
aldehydodiketones, **IE**
alditols, **IF**
aldobiuronic acids, **IF**
aldosterone, **IID**
alginic acid, **IFG** supplement
alkanes, **IA**
alkenes, diphenyl, **IIIF**
alkanes, diphenyl, **IIIF**
alkanes, tetraphenyl, **IIIF**
alkanes, triphenyl, **IIIF**
alkenes, **IA**, cyclic **IIA, IIB, IE** supplement
alkynes, **IA, IE** supplement
alkynes, aromatic, **IIIE**
alkynes, diphenyl, **IIIF**
allenes, **IE** supplement, cyclic **IIA, IIB**
allenes, diphenyl, **IIIF**
allopregnane compounds, **IID**
alloxazines, **IVL**
aluminium compounds, **IVK, IB**
aluminium compounds, aromatic, **IIIB**
amaryllidaceae alkaloids, **IVB, IVL**
amides, **IC**, cyclic **IIA, IIB**
amides, aryl, **IIIB**
amides, aromatic, **IIIG**
amidines, **IB**
amidoximes, **IC**
amines, **IB**, cyclic **IIA, IIB**
amines, aromatic, **IIIB**
amines, aralkyl, **IIID**
amino acids, **ID**
amino-sugars, **IF**
aminoaralkanols, **IIIE**
aminocyclitols, **IIAB** supplement
aminodicarboxylic acids, **IE**
aminophenols, **IIIA**
andrenocortical hormones, **IID**
androgens, **IID**

androstane compounds, **IID**
anhydrides, **IC**
aniline, **IIIB**
aniline black, **IVIJ**
anthracene, **IIIH**
anthramycin, **IVL**
anthranilic acid, **IIIG, IVG**
anthraquinones, **IIIH**
antimony compounds, **IB, IVG, IVK**
antimony compounds, aromatic, **IIIB**
aporphine alkaloids, **IVL**
aporphines, **IVG, IVH**
aporphinoids, **IVH**
aposafranines, **IVIJ**
arabans, **IFG** supplement
aralkanediols, **IIIE**
aralkanes, dicarbonyl, **IIIE**
aralkanetriols, **IIIE**
aralkanols, **IIID**
aralkanols, dihydric, **IIIE**
aralkenes, **IIIE**
aralkylamines, **IIID**
aranotins, **IVL**
arene acids, **IIIE**
arene carboxylic acids, **IIIG**
arenediazonium compounds, **IIIC**
arenes, acyl, **IIIE**
arenes, unsaturated side-chain, **IIIE**
areneselenols, **IIIA**
aromatic substitution, electrophilic, **IIIA**
aromatic substitution, homolytic, **IIIA**
aromatic substitution, nucleophilic, **IIIA**
arphamenines, **IVL**
arsenic compounds, **IVG, IVK**
arsenic compounds, aromatic, **IIIB**
arsenic compounds, **IB**
aryl azides, **IIIC**
aryl metal complexes, **IIIB**
arylamides, **IIIB**
arylhydroxylamines, **IIIB**
arylnitramines, **IIIC**
aryltellurium compounds, **IIIA**
asukamycin, **IVL**

atisine alkaloids, **IVG**
atisirane group, **IIC**
atranes, **IVK**
atropic acids, **IIIE**
azaborines, **IVK**
azadiborines, **IVK**
azadisilines, **IVK**
azaindoles, **IVH**
azalumines, **IVK**
azamycins, **IVL**
azaphosphorines, **IVK**
azaporphins, **IVB**
azasilines, **IVK**
azepine, **IVH**
azepins, **IVK**
azetidine-2-carboxylic acid, **IVL**
azetidines, **IVA**
azides, aryl, **IIIC**
azidobenzoic acid, **IIIG**
aziridines, **IVA**
azirines, **IVA**
azo-arenes, **IIIC**
azocine, **IVK**
azoximes, **IVD**
azoxy compounds, aromatic, **IIIC**
azoxyphenols, **IIIA**
azulene group, **IIC**
azulenes, **IIIG**

B
banzodiazepine bases, **IVL**
barium compounds, **IB**
barium compounds, aromatic, **IIIB**
benzo[c]pyran, **IVE**
benazaulenes, **IIIH**
benzazepins, **IVK**
benzacridines, **IVG**
benzamidates, **IIIG**
benzanthracenes, **IIIH**
benza[a]anthracene, **IIIH**
benza[b]anthracene, **IIIH**
benzene, **IIIA**
benzene, alkyl, **IIIE**

benzene, halogenated, **IIIA**
benzene, polycarboxylic acids, **IIIE**
benzene, polyhydroxy-, **IIIA**
benzilic acids, **IIIF**
benzimidazoles, **IVC**
benzindan group, **IIIH**
benzisoquinolines, **IVG**
benzisothiazoles, **IVC**
benzisoxazoles, **IVC**
benzochrysenes, **IIIH**
benzocinnolines, **IVIJ**
benzocyclobutene, **IIIG**
benzocyclopropene, **IIIG**
benzodiazepin bases, **IVL**
benzofurazans, **IVD**
benzoic acid, **IIIG**
benzoic acids, acyl, **IIIG**
benzoic acids, dihydroxy, **IIID**
benzoic acids, hydroxy, **IIID**
benzoic acids, trihydroxy, **IIID**
benzoindoles, **IVA**
benzoins, **IIIF**
benzophenanthridines, **IVG**
benzoporphins, **IVB**
benzopyrroles, **IVA**
benzopyrazines, **IVIJ**
benzopyrenes, **IIIH**
benzopyrrocolines, **IVG**
benzoquinols, **IIIB**
benzoquinones, **IIIB, IIAB** supplement
benzoselenazoles, **IVC**
benzotellurophenes, **IVA**
benzothiadiazoles, **IVD**
benzothiazines, **IVH**
benzothiazoles, **IVC**
benzothiopyrans, **IVE**
benzothioxanthene, **IVE**
benzotriazoles, **IVD**
benzotropones, **IIIG**
benzoxathiins, **IVH**
benzoxazines, **IVH**
benzoxazinones, **IVL**
benzoxazoles, **IVC**

benzoxocins, **IVK**
benzo[b]furans, **IVA**
benzo[b]pyrans, **IVE**
benzo[b]selenophenes, **IVA**
benzo[b]thiophene, **IVA**
benzo[b]thiopyrans, **IVE**
benzo[c]thiophenes, **IVA**
benzo[c]thiopyran, **IVE**
benzo[f]quinolines, **IVG**
benzo[g]quinolines, **IVG**
benzo[h]quinolines, **IVG**
benzpinacols, **IIIF**
benzylisoquinolines, **IVG**
benzynequinone, **IIIB**
berberines, **IVG**
beryllium compounds, **IB**
beryllium compounds, aromatic, **IIIB**
biacridines, **IVG**
bianthraquinones, **IIIH**
bianthryls, **IIIH**
biazaluminines, **IVK**
biflavanones, **IVE**
biindenyl, **IIIG**
bile acids, **IID**
bile pigments, **IVB**
binaphthyls, **IIIG**
biphenyl, **IIIF**
bipiperidyls, **IVF**
bipyrrolyls, **IVA**
biquinolines, **IVF**
bisabolenes, **IIC**
bisbenzylisoquinoline alkaloids, **IVL**
bisbenzylisoquinolines, **IVG, IVH**
bisindole alkaloids, **IVB**
bismuth compounds, **IB, IVG**
bismuth compounds, aromatic, **IIIB**
boranes, **IB**
boron compounds, **IB, IVK**
boron compounds, aromatic, **IIIB**
boxburghines, **IVB**
brazilin group, **IVE**
brevianamides, **IVB, IVL**
bridged ring systems, **IIC**

bromides, alkyl, **IA**, cyclic **IIA, IIB**
butane, cyclo-, **IIA**
buxus alkaloids, **IVG**

C
cadinenes, **IIC**
cadinols, **IIC**
cadmium compounds, **IB**
cadmium compounds, aromatic, **IIIB**
cesium compounds, **IB**
cesium compounds, aromatic, **IIIB**
calarane group, **IIC**
calcium compounds, **IB**
calcium compounds, aromatic, **IIIB**
calycanthidine, **IVG**
calycanthine, **IVG**
cadmium compounds, **IB, IVL**
camphor, **IIC**
capsaicin, **IVL**
captothecin, **IVG**
carane group, **IIC**
carbamates, **IC**
carbapenem antibiotics, **IVL**
carbazole, **IVA**
carbenes, **IE** supplement, cyclic **IIA, IIB**
carbohydrazides, **IC**
carboline alkaloids, **IVL**
carboline, **IVB, IVH**
carbon monoxide, **IC**
carbonic acid, **IC**
carbonyl compounds, di, **ID**
carboxylic acids, **IC**, cyclic **IIA, IIB**
carboxylic acids, amino, **ID**
carboxylic acids, arene, **IIIG**
carboxylic acids, di-, **ID**
carboxylic acids, di-, amino, **IE**
carboxylic acids, di-, formyl, **IE**
carboxylic acids, di-, hydroxy, **IE**
carboxylic acids, di-, hydroxyoxo, **IG**
carboxylic acids, di-, oxo, **IE**
carboxylic acids, dihydroxy, **IE**
carboxylic acids, dioxo-, **IE**
carboxylic acids, hydroxyoxo, **IE**

carboxylic acids, nitro, **ID**
carboxylic acids, tetra-, **IG**
carboxylic acids, thio, **IC**
carboxylic acids, tri-, **IE**
carboxylic acids, tri-, hydroxy, **IG**
carboxylic acids, trihydroxy, **IG**
carotenoids, **IIB**
carrageenan, **IFG** supplement
caryophyllene, **IIC**
cedrol, **IIC**
cembrane group, **IIC**
cephalosporins, **IVL**
chalcone, **IIAB** supplement, **IIIF**
chimonanthine, **IVG**
chloramphenicol, **IVL**
chlorides, alkyl, **IA**, cyclic **IIA**, **IIB**
chlorophyll, **IVB**
cholanoic acids, **IID**
cholesterol, **IID**
chondroitin sulphates, **IFG** supplement
chroman, **IVE**
chromanols, **IVE**
chrysenes, **IIIH**
chrysenes, benzo-, **IIIH**
cinnamic acid, **IIIE**
cinnolinecarboxylic acids, **IVIJ**
cinnolines, **IVIJ**
cinodine, **IVL**
clavulanic acid, **IVL**
coccinine, **IVB**
coclaurine, **IVL**
colchicine, **IVG**, **IVH**
condensed cyclic systems, **IIC**
congressane, **IIC**
coronatine, **IVL**
corroles, **IVB**
coumarin, **IIAB**
cryptolepine, **IVG**
cularines, **IVG**, **IVH**, **IVL**
cummulenes, **IE** supplement
curcumenes, **IIC**
cyanocobalamin, **IVB**
cycanines, **IVB**

cyclazines, **IVH**
cyclic, poly- compounds, **IIC**
cyclic systems, bridged, **IIC**
cyclic systems, condensed, **IIC**
cyclic systems, fused, **IIC**
cyclitols, **IIAB** supplement
cyclopiazonic acid, **IVL**
cyclobutene, benzo-, **IIIG**
cycloheptane, **IIB**
cyclohexadiene, **IIAB** supplement
cyclohexadienones, hydroxy, **IIIB**
cyclohexane, **IIB**
cycloheximide, **IVL**
cyclopentadiene, **IIAB** supplement
cyclopentadienyl, metal, **IIAB** supplement
cyclopentane, **IIA**
cyclophanes, caged, **IIIF**
cyclophanes, layered, **IIIF**
cyclophet[fg]acenaphthylene, **IIIH**
cyclopropane, **IIA**
cyclopropene, benzo-, **IIIG**
cylobutane, **IIA**
cytochalasins, **IVL**

D
damascenine, **IVL**
deoxy-sugars, **IF**
depsidones, **IIID**
depsones, **IIID**
dermatan sulphate, **IFG** supplement
diacids, **ID**
dialdehydes, **ID**
dialdehydes, dihydroxy, **IG**
dialdehydes, hydroxy, **IE**
dialdehydoketones, **IE**
diazaborines, **IVK**
diazadialuminines, **IVK**
diazanaphthalenes, **IVH**
diazaphosphorines, **IVK**
diazasilines, **IVK**
diazepines, **IVK**
diazetidines, **IVA**
diazetines, **IVA**

diazines, **IVIJ**
diaziridines, **IVA**
diazo compounds, aromatic, **IIIC**
diazoamino compounds, aromatic, **IIIC**
diazocines, **IVK**
diazonium compounds, aromatic, **IIIC**
diazophenols, **IIIA**
dibenzanthracenes, **IIIH**
dibenzazepines, **IVK**
dibenzophospholes, **IVA**
dibenzopyrazines, **IVIJ**
dibenzopyrrocolline alkaloids, **IVH**
dibenzothiophenes, **IVA**
dibenzothiopyran, **IVE**
dicarboxylic acids, **ID**
dicarboxylic acids, amino, **IE**
dicarboxylic acids, aromatic, **IIIE**
dicarboxylic acids, formyl, **IE**
dicarboxylic acids, hydroxy, **IE**
dicarboxylic acids, oxo, **IE**
dienes, **IE** supplement, cyclic **IIA, IIB**
dienes, aromatic, **IIIE**
diglycerides, **IE** supplement
dihydric phenols, **IIIA**
dihydrofurans, **IVA**
dihydroxyaldehydes, **IE**
dihydroxycarboxylic acids, **IE**
dihydroxyketones, **IE**
diketones, dihydroxy, **IG**
diketones, aldehydo, **IE**
dimethyltryptamine, **IVL**
dioxaborins, **IVK**
dioxadiborins, **IVK**
dioxadigermalins, **IVK**
dioxadisilins, **IVK**
dioxagermasilins, **IVK**
dioxanes, **IVH**
dioxaphosphorins, **IVK**
dioxarsenins, **IVK**
dioxathianes, **IVIJ**
dioxathiazoles, **IVD**
dioxatimonins, **IVK**
dioxazines, **IVIJ**

dioxazoles, **IVD**
dioxepins, **IVK**
dioxetanes, **IVA**
dioxocarboxylic acids, **IE**
diphenylmethane, **IIIF**
dipsides, **IIID**
disaccharides, **IF**
diselenazoles, **IVD**
diterpenoids, **IIC**
dithanes, **IVH**
dithiadiazines, **IVIJ**
dithiadiazoles, **IVD**
dithiazepines, **IVK**
dithiazines, **IVIJ**
dithiazoles, **IVD**
dithietanes, **IVA**
dithietenes, **IVA**
dithiins, **IVH**
diketones, hydroxy, **IE**
dolichotheline, **IVL**

E
eburicoic acid, **IIC**
echinine, **IVG**
echinorine, **IVG, IVL**
echinulin, **IVL, IVB**
elaiomycin, **IVL**
electrophilic aromatic substitution, **IIIA**
elemane, **IIC**
enol, lactones, **ID**
enynes, **IE supplement**
eremophilene, **IIC**
ergosterol, **IID**
ergot alkaloids, **IVB, IVL**
erythrina alkaloids, **IVL**
esters, **IB, IC,** cyclic **IIA, IIB**
esters, aromatic, **IIIG**
esters, keto, **ID**
estrone, **IID**
ethanophenanthridine alkaloids, **IVB**
ethers, **IB**
ethylene oxides, **IVA**
ethylene sulphides, **IVA**

ethyleneimines, **IVA**
eudalene, **IIC**
eurhodines, **IVIJ**
eurhodols, **IVIJ**
evodiamine, **IVL**
extracellular polysaccharides, **IFG** suplmt

F
farnesene, **IIC**
flavanones, **IVE**
flavans, **IVE**
flavins, **IVL**
flavonoids, **IIAB** supplement
fluoranthene group, **IIIH**
fluorene group, **IIIH**
fluorides, alkyl, **IA**
fluoridines, **IVIJ**
folicanthine, **IVL**
folicanthine, **IVG**
formates, **IC**
formyldicarboxylic acids, **IE**
fucan, **IFG** supplement
fucogalactoxyloglucans, **IFG** supplement
fulminic acid , **IC**
fulvenes, **IIAB** supplement
fumaric acid, **ID**
furamones, **IE** supplement
furan, **IVA**
furans, dihydro, **IVA**
furans, tetrahydro, **IVA**
furopyridines, **IVH**
furoquinolines, **IVG, IVH, IVL**
fused cyclic systems, **IIC**

G
galactoglucomannans, **IFG** supplement
galactomannans, **IFG** supplement
galanthamine alkaloids, **IVB**
gallium compounds, **IB**
gallium compounds, aromatic, **IIIB**
gallocyanines, **IVIJ**
geldanomycin, **IVL**

germacranolides, **IIC**
germanium compounds, **IB, IVE, IVK**
germanium compounds, aromatic, **IIIB**
gestogens, **IID**
gibberellic acid, **IIC**
gliotoxin, **IVL**
glucomannans, **IFG** supplement
glutaric acid, **ID**
glycerides, **IE, IE** supplement
glyceritol, **IE**
glycerol, **IE, IE** supplement
glycolipids, **IE, IF**
glycols, **ID**
glycoproteins, **IF, IFG** supplement
glycosides, aglycons, **IID**
glycosides, cardiotonic, **IID**
gramine, **IVL**
graveoline, **IVL**
Grignard reagents, **IB**
guaiol, **IIC**
guanidine, IC

H
haem, **IVB**
haematoxylin group, **IVE**
halides, alkyl, **IA**, cyclic **IIA, IIB**
halides, alkene, **IA**, cyclic **IIA, IIB**
halides, alkyne, **IA**
halobenzene, **IIIA**
hasubanonine, **IVL**
hasubanonine group, **IVG**
hemicellulose, **IFG** supplement
hemicyanines, **IVB**
hemoglobin, **IVB**
heparan sulphate, **IFG** supplement
heparin, **IFG** supplement
heptane, cyclo-, **IIB**
heptitols, **IF**
hexane, cyclo-, **IIB**
hexitols, **IF**
histamine, **IVC**
homoadamantane, **IIC**
homolytic aromatic substitution, **IIIA**

homomorphine alkaloids, **IVG**
humic acid, **IIID**
humulenes, **IIC**
hyaluronic acid, **IFG** supplement
hydrazides, aromatic, **IIIG**
hydrazidines, **IC**
hydrazine derivatives, aromatic, **IIIC**
hydrazonates, aromatic, **IIIG**
hydrazophenols, **IIIA**
hydrobenzofurans, **IVA**
hydrobenzoins, **IIIF**
hydrocarbaxoles, **IVA**
hydroindoles, **IVA**
hydroisobenzofurans, **IVA**
hydropyrans, **IVE**
hydrotellurophenes, **IVA**
hydroxamic acids, **IC**
hydroxamic acid -oximes, **IC**
hydroximoyl chlorides, **IC**
hydroxy ketoaldehydes, **IE**
hydroxy acids, **ID**
hydroxyaryl carbaldehydes, **IIID**
hydroxybenzoic acids, **IIID**
hydroxycyclohexadienones, **IIIB**
hydroxydialdkeydes, **IE**
hydroxydicarboxylic acids, **IE**
hydroxydiketones, **IE**
hydroxyflavanones, **IVE**
hydroxylamines, **IIIG**
hydroxylamines, aryl, **IIIB**
hydroxyoxocarboxylic acids, **IE**
hydroxyphenyl ketones, **IIID**
hydroxytriones, **IG**

I
ilicicolin H, **IVL**
imadazole, **IVC**
imidates, alkyl **IC**
imidazole alkaloids, **IVC**
imidazolidines, **IVC**
imidazolo[4,5-d]pyrimidines, **IVL**
imines, benzoquinone, **IIIB**
indan, **IIIG**

indanobenzazepines, **IVG**
indazole, **IVC**
indene, **IIIG**
indigotin, **IVB**
indigo, **IVB**
indirubin, **IVB**
indium compounds, **IB**
indium compounds, aromatic, **IIIB**
indole alkaloids, **IVL**
indoles, **IVA**
indolines, **IVA**
indolinones, **IVA**
indolizines, **IVH**
indolmycin, **IVL**
indoloquinazoline alkaloids, **IVIJ**
indulines, **IVIJ**
inidazoline, **IVC**
inositols, **IIB,**
iodides, alkyl, **IA**, cyclic **IIA, IIB**
iodine compounds, **IVK**
ipecac alkaloids, **IVL**
ipecacuanha alkaloids, **IVG, IVH**
isobenzofurans, **IVA**
isochromene, **IVE**
isocyanides, **IC**
isofebrifugine, **IVIJ**
isoflavanones, **IVE**
isoflavans, **IVE**
isoindigo, **IVB**
isoindoles, **IVA**
isoindolines, **IVA**
isonitriles, **IVL**
isopavinanes, **IVH**
isopavines, **IVG**
isophthalic acid, **IIIE**
isoprenoid indole bases, **IVL**
isoquinoline alkaloids, **IVF, IVH, IVL**
isothiazoles, **IVC**
isothiochromeme, **IVE**
isochroman, **IVE**
isoxazole, **IVC**

J
Jaborandi alkaloids, **IVC**
julolidine, **IVH**

K
kaurane group, **IIC**
keratan sulphate, **IFG** supplement
ketals, **IE**
ketenes, diphenyl, **IIIF**
keto acids, **ID**
ketoaldehydes, **ID**
ketoaldehydes, hydroxy, **IE**
ketones, **IC**
ketones, aromatic, **IIID**
ketones, dialdehydo, **IE**
ketones, di-, aldehydo, **IE**
ketones, di-, dihydroxy, **IG**
ketones, dihydroxy, **IE**
ketones, di-, hydroxy, **IE**
ketones, hydroxyphenyl, **IIID**
ketones, tri-, **IE**
ketones, unsaturated, aromatic, **IIIE**
keto esters, **ID**
kitol, **IIB**

L
labdanes, **IIC**
lactam antibiotics, **IVL**
lactams, **ID**
lactams, aromatic, **IIIE**
lactones, **ID**
lactones, aromatic, **IIIE**
lactones, enol, **ID**
lanosterol, **IIC**
lead compounds, **IB, IVK**
lead compounds, aromatic, **IIIB**
lignans, **IIID**
lignin, **IIID**
lilolidine, **IVH**
lincomycin, **IVL**
lipids, **IE**
lipopolysaccharides, **IFG** supplement
lithium compounds, **IB**

lithium compounds, aromatic, **IIIB**
lnositols, **IIAB** supplement
lupeol, **IIC**
lupinane alkaloids, **IVH**
lycocytonine alkaloids, **IVG**
lythraceae alkaloids, **IVH**

M
macrocycles, **IIB**
magnesium compounds, **IB, IVK**
magnesium compounds, aromatic, **IIIB**
maleic acid , **ID**
malonic acid , **ID**
malonimides, **IVA**
malonomicin, **IVL**
mandelic acid, **IIIE**
manthine, **IVB**
mechanisms, **IA**
menthane, **IIB**
mercury compounds, **IB, IVK**
mercury compounds, aromatic, **IIIB**
mesembrine alkaloids, **IVL**
mesochimonanthine, **IVG**
metacyclophanes, **IIIF**
metal arene complexes, **IIIB**
metal complexes, aryl, **IIIB**
metallocenes, **IIAB** supplement
methane, diphenyl-, **IIIF**
methane, triphenyl, **IIIF**
methylgranatanine alkaloids, **IVB**
metomycins, **IVL**
mitomycins, **IVL**
monosaccharides, synthesis with, **IEFG** supplement
monosaccharides, **IF**
montanine, **IVB**
morphinan alkaloids, **IVL**
morphine alkaloids, **IVG**
morphine-thebaine group, **IVG**
mucohalic acids, **IE** supplement
mutilin, **IIC**
mycelianamide, **IVL**
myxopyronin A, **IVL**

N

Nametkin rearrangement, **IIC**
nanadrides, **IIAB** supplement
naphthacene, **IIIH**
naphthalene, **IIIG**
naphthoisoquinolines, **IVG**
naphthols, **IIIG**
naphthopyrans, **IVE**
naphthoquinolines, **IVG**
naphthoquinones, **IIIG**
naphthyridines, **IVH**
naphthyridinomycin, **IVL**
necines, **IVB**
nicic acids, **IVB**
nicotine, **IVL**
nigrosines, **IVIJ**
nitramines, aryl, **IIIC**
nitrile oxides, aromatic, **IIIG**
nitriles, **IC**, cyclic **IIA, IIB**
nitriles, aromatic, **IIIG**
nitro compounds, **IB**, cyclic **IIA, IIB**
nitro acids, **ID**
nitro derivatives, aromatic, **IIIB**
nitrophenols, **IIIA**
nitroso compounds, aromatic, **IIIC**
nitroso deriveratives, aromatic, **IIIB**
nocardicins, **IVL**
norborane group, **IIC**
norlaudanosoline synthase, **IVL**
norsteroids, **IID**
nucleic acids, **IVL**
nucleophilic aromatic substitution, **IIIA**
nucleosides, **IVL**
nucleotides, **IVL**
nuphar alkaloids, **IVH**
nybomycin, **IVL**

O

oestrogens, **IID**
olefins, **IA**, cyclic **IIA, IIB**
oligosaccharides, **IF**
organometallic compounds, **IB**
ormosia alkaloids, **IVH**

osazones, aromatic, **IIIC**
oxaborins, **IVK**
oxadiazines, **IVIJ**
oxadiaziridine, **IVA**
oxadiazoles, **IVD**
oxadiphosphorinanes, **IVK**
oxadithianes, **IVIJ**
oxalic acid, **ID**
oxaphosphorins, **IVK**
oxarsenins, **IVK**
oxathiadiazines, **IVIJ**
oxathianes, **IVH**
oxathiazines, **IVIJ**
oxathiazoles, **IVD**
oxathietanes, **IVA**
oxathiins, **IVH**
oxatriazines, **IVIJ**
oxatriazoles, **IVD**
oxazepines, **IVK**
oxazine, **IVIJ**, **IVH**
oxaziranes, **IVA**
oxaziridines, **IVA**
oxazoles, **IVC**
oxepins, **IVK**
oxetanes, **IVA**
oxetanols, **IVA**
oxetanones, **IVA**
oxetenes, **IVA**
oxetes, **IVA**
oxirans, **IVA**
oxirenes, **IVA**
oxoaporphines, **IVH**
oxocin, **IVK**
oxodicarboxylic acids, **IE**
oxoindolines, **IVA**
oxonins, **IVK**
oxonol dyes, **IVB**

P
PAF, **IEFG** supplement
papaverine, **IVL**
papaverrubines, **IVH**
paracyclophane, **IIIF**

patchouli alcohol, **IIC**
pavinanes, **IVH**
pavines, **IVG**
pectic substances, **IFG** supplement
penicillins, **IVL**
pentacene, **IIIH**
pentane, cyclo-, **IIA**
pentaphene, **IIIH**
pentazines, **IVIJ**
pentitols, **IF**
peptides, **ID**
peptidoglycan, **IFG** supplement
peracids, **IC**, cyclic **IIA, IIB**
pericyclic reactions, **IIAB** supplement
perinaphthothiapyrans, **IVE**
peroxides, alkyl **IB** , acid **IC**
perylene, **IIIH**
phenalene, **IIIG**
phenanthrene, **IIIH**
phenanthridine, **IVG**
phenanthroindolizine alkaloids, **IVL**
phenanthrolines, **IVH**
phenazine, **IVIJ, IVL**
phenazinols, **IVIJ**
phenethylamines, **IVL**
phenethylisoquinoline alkaloids, **IVL**
phenethylisoquinolines, **IVH**
phenol, **IIIA**
phenol, amino-, **IIIA**
phenol, aralkylamine, **IIID**
phenol, dihydric, **IIIA**
phenol, nitro-, **IIIA**
phenolsulphonic acids, **IIIA**
phenothiazine dyes, **IVIJ**
phenothiazines, **IVH**
phenoxazine dyes, **IVIJ**
phenoxazines, **IVH**
phenoxazinones, **IVL**
phenylacetylenes, **IIIE**
phenylalanine, **IVL**
phenylchromans, **IVE**
phenylethylisoquinolines, **IVG**
phomazarin, **IVL**

phospholes, **IVA**
phospholipase A2 inhibitors, **IEFG** supplement
phospholipids, **IE**
phosphorus compounds, **IB, IVG, IVK**
phosphorus compounds, aromatic, **IIIB**
phthalazines, **IVIJ**
phthalideisoquinoline alkaloids, **IVH**
phthalideisoquinolines, **IVG**
phthalides, **IIIE**
phthalocyanines, **IVB**
piane group, **IIC**
piazthioles, **IVD**
pimaranes, **IIC**
piperidine alkaloids, **IVG, IVL**
platelet-activating factor, **IEFG** supplement
pleiadene, **IIIH**
polonium compounds, **IB**
polycarbocyclic compounds, **IIC**
polyenes, **IE** supplement
polyhydroxybenzenes, **IIIA**
polypeptides, **ID**
polysaccharides, **IF**
porphin, **IVB**
porphyrinpropionic acids, **IVB**
porphyrins, **IVB**
potassium compounds, **IB**
potassium compounds, aromatic, **IIIB**
pregnene compounds, **IID**
prescamines, **IVB**
pretazettine alkaloids, **IVB**
proaporphines, **IVH**
prodiginines, **IVL**
prodigiosin, **IVL**
propane, cyclo-, **IIA**
prostaglandins, **IIAB** supplement
proteoglycans, **IFG** supplement
protoberberine, **IVL, IVH**
protopine alkaloids, **IVH**
protopines, **IVG**
protostephanine, **IVL**
pseudanes, **IVG**
pseudans, **IVL**
pseudosugars, **IIAB** supplement

pseurotins, **IVL**
pteridines, **IVL**
pulvinic acid, **IE** supplement
purines, **IVL**
pyracene, **IIIH**
pyran, **IVE**
pyranopyridines, **IVH**
pyranoquinolines, **IVG**
pyrazines, **IVIJ**
pyrazol-l-ylaline, **IVL**
pyrazole, **IVC**
pyrene, **IIIH**
pyrene, benzo-, **IIIH**
pyridazines, **IVIJ**
pyridindoles, **IVH**
pyridine, **IVF**
pyridine alkaloids, **IVG, IVL**
pyrimidines, **IVIJ**
pyrroles, **IVA**
pyrrolidines, **IVA, IVL**
pyrrolines, **IVA**
pyrrolizidine alkaloids, **IVL**
pyrrolnitrin, **IVL**
pyrrolopyridines, **IVH**
pyrromethenes, **IVB**

Q
quanazoline alkaloids, **IVIJ**
quantitative analysis, **IA**
quercitols, **IIB**
quinazolines, **IVIJ**
quinazoline alkaloids, **IVL**
quinoline, **IVF**
quinoline alkaloids, **IVG, IVL**
quinolines, **IVL**
quinoloizines, **IVH**
quinols, **IIAB** supplement
quinoxalines, **IVIJ, IVL**
quinuclidine, **IVH**

R
radicals, **IA,** cyclic **IIA, IIB**
reaction mechanisms, **IA**

reticuline, **IVL**
retinene, **IIB**
rhoeadines, **IVG, IVH**
rifamycins, **IVL**
rosenonolactone, **IIC**
rubidium compounds, **IB**
rubidium compounds, aromatic, **IIIB**
rubrene, **IIIH**
rutaecarpine, **IVL**

S

saframycin A, **IVL**
safranines, **IVIJ**
salamander alkaloids, **IVG**
salicyclic acids, **IIIG**
santalenes, **IIC**
santene group, **IIC**
sapogenins, steroid, **IIE**
saponins, steroid, **IIE**
sapphyrins, **IVB**
seaweed mucilages, **IFG** supplement
secamines, **IVB**
selenadiazoles, **IVD**
selenazoles, **IVC**
selenium compounds, **IB**
selenium compounds, aromatic, **IIIA**
selenophenes, **IVA**
selenopyrans, **IVE**
sesquimeric compounds, **IVB**
sesquiterpenes, **IIAB** supplement
sesquiterpenoids, **IIC**
sesterterpenoids, **IIC**
shihunine, **IVL**
shikimic acid, **IIAB** supplement
sibiromycin, **IVL**
silacyclopentadienes, **IVA**
silicon compounds, **IB, IVE, IVK**
silicon compounds, aromatic, **IIIB**
sinomenine-salutaridine group, **IVG**
sirodesmin, **IVL**
sodium compounds, **IB**
sodium compounds, aromatic, **IIIB**

soyasapogenol D, **IIC**
spectroscopy, **IA**
sphingolipids, **IE**
spiro compounds, **IIC**
spirobenzylisoquinolines, **IVG**
sporidesmin, **IVL**
squalene, **IIC**
stachane group, **IIC**
stereochemistry, **IA**
steroid sapogenins, **IIE**
steroid saponins, **IIE**
steroidal alkaloids, **IVL**
steroids, **IID, IIE**
steroids, biogenesis, **IIE**
sterols, **IID**
stigmasterol, **IID**
stilbenes, **IIIF**
streptothricin, **IVL**
streptovaricins, **IVL**
strontium compounds, **IB**
strontium compounds, aromatic, **IIIB**
strophanthidin, **IID**
styrene, **IIIE**
styryl, **IVB**
succinic acid, **ID, IE**
sugars, amino, **IF**
sugars, deoxy-, **IF**
sugars, pseudo-, **IIAB** supplement
sugars, thio-, **IF**
sulfur compounds, **IB**, cyclic **IIA, IIB**
sulphenic acids, **IB**
sulphenyl compounds, aromatic, **IIIA**
sulphides, aromatic, **IIIA**
sulphides, **IB**
sulphinic acids, alkyl **IB**
sulphinic acids, aromatic, **IIIA**
sulphones, **IB**
sulphonic acids, **IB**
sulphonic acids, aromatic, **IIIA**
sulphonium compounds, **IB**
sulphoxides, **IB**

T

tannins, **IIID**
teichoic acids, **IF**
teichoic acids, **IFG** supplement
tellurium compounds, **IB, IIIA**
tellurophenes, **IVA**
telluropyrans, **IVE**
tenellin, **IVL**
tenuazonic acid, **IVL**
terephthalic acid, **IIIE**
terpenes, biogenesis, **IIE**
terpenes, monocyclic, **IIB**
terpenes, mono-, acyclic, **IIB**
terpenes, mono-, heterocyclic, **IIB**
terpenoid alkaloids, **IVL**
terphenyl, **IIIF**
tetra-ones, **IG**
tetracene, **IIIH**
tetrahydrofurans, **IVA**
tetraphene, **IIIH**
tetrasaccharides, **IF**
tetrasilacyclohexanes, **IVK**
tetrathianes, **IVIJ**
tetrazenes, **IIIC**
tetrazines, **IVIJ**
tetrazocine, **IVK**
tetrazoles, **IVD**
tetritols, **IG**
tetronic acids, **IE** supplement
thallium compounds, **IB**
thallium compounds, aromatic, **IIIB**
thiabicycloalkanes, **IVE**
thiaborins, **IVK**
thiadiazepines, **IVK**
thiadiazines, **IVIJ**
thiadiazoles, **IVD**
thiaselenazoles, **IVD**
thiatellurazoles, **IVD**
thiatriazines, **IVIJ**
thiatriazoles, **IVD**
thiazepines, **IVK**
thiazine, **IVIJ, IVH**
thiazoles, **IVC**

thiepin, **IVK**
thietanes, **IVA**
thiiranes, **IVA**
thio-sugars, **IF**
thiocarboxylic acids, **IC**
thiochromenes, **IVE**
thiolanes, **IVA**
thiolenes, **IVA**
thiols, **IB**, cyclic **IIA, IIB**
thiophene, **IVA**
thiophenol, **IIIA**
thiopyrans, **IVE**
thiovulpinic acids, **IE** supplement
thioxanthene, **IVE**
thujane group, **IIC**
tin compounds, **IB, IVE, IVK**
tin compounds, aromatic, **IIIB**
toad poisons, **IID**
tocopherols, **IVE**
tomaymycin, **IVL**
trachylobane group, **IIC**
transition metals, **IB**
triamantane, **IIC**
triazaborines, **IVK**
triazasilines, **IVK**
triazene oxides, **IIIC**
triazines, **IVIJ**
triazocines, **IVK**
triazoles, **IVD**
tricarboxylic acids, **IE**
triglycerides, **IE** supplement
triketones, **IE**
trimethylene sulphides, **IVA**
trimethylene oxides, **IVA**
trioxadiborinanes, **IVK**
trioxanes, **IVIJ**
trioxepins, **IVK**
triphendioxazines, **IVIJ**
triphenylene, **IIIH**
triphenylmethane, **IIIF**
triptycene, **IIIH**
trisaccharides, **IF**
trisilacyclohexanes, **IVK**

triterpenoids, **IIC**
trithianes, **IVIJ**
trizadiborines, **IVK**
tropane, **IVL**
tropane alkaloids, **IVB**
tropone, **IIB**
tropylium complexes, **IIIB**
tropylium salts, **IIB**
tryptophan, **IVG**
tuberin, **IVL**
twistane, **IIC**
tyrosine metabolites, halogenated, **IIAB**
tyrosine, **IVL**

U
urea, **IC**

V
valerenic acid, **IIC**
vasicine alkaloids, **IVIJ**
vebrifugine, **IBIJ**
virginiamicin antibiotics, **IVL**
viridicatine, **IVG**
viridicatol, **IVG**
vitamin A, **IIB**

W
Wagner-Meerwein rearrangement, **IIC**
widdrol, **IIC**

X
xanthene, **IVE**
xanthophylls, **IIB**
xylans, **IFG** supplement

Z
zinc compounds, **IB**
zinc compounds, aromatic, **IIIB**
zingiberenes, **IIC**

B.32.

Science Citation Index:

Q1.S19

- citations of papers, patents, or books
- published titles, authors, and institutions
- review articles

Science Citation Index is a compilation of all scientific publications in a particular year that have cited a given paper, patent, or book (key reference). It enables one to search forward from a key reference.

EASY ACCESS

— To find all citations of a key reference, look up the name of the first author in **Science Citation Index**.
— To identify all papers from a particular institution, look up the organization in the Corporate Index.
— To find articles on a particular subject, look up the subject in the Permuterm Subject Index.

INDEXING

The **Citation Index** is an alphabetical list of references given in bibliographies and footnotes of source articles arranged by *first* author. Under the name of each cited author, cited items are arranged chronologically, and within years, alphabetical by journal title abbreviation. Those references written by anonymous authors follow the letter Z in the Citation Index and are arranged alphabetically by journal title abbreviation, and then chronologically.

The **Patent Citation Index** is a numerical listing of cited patents by number devoid of alpha characters. The patent reference year, inventor, country, and application or reissue status is given under the patent number. The citing items follow, arranged alphabetically first by author and then by journal title abbreviation.

The **Permuterm Subject Index** is an alphabetical subject index derived from key words in the titles of source articles. Every significant word in a title is paired with the other key words from that title. These pairs are permuted so that each term in the pair appears both as a primary and a secondary term. Every entry references the first author of a source article.

The **Source Index** is an alphabetical listing of all the authors published during the listing period. Under the primary author's name, co-authors, article title, and reference information are listed in alphabetical order by journal title abbreviation. All co-authors are cross-referenced to the first authors.

The **Corporate Index** is divided into: (1) the Geographic section lists source items alphabetically by country and city of the author's organization. The names of states are given when the country is the United States. The states are listed first, followed by the nations of the world.

(2) The Organization section is an alphabetical list of source authors' organizational affiliations, giving the geographic location for each organization.

COMPUTER SEARCH

Science Citation Index is available through Dialog Information Retrieval Service (Database [34,434]) or through STN International (FILE SCISEARCH). The Basic Index (BI) contains single words from the title (TI), author key words (ST), abstract (AB), and Key Words Plus (STP) fields. Example search fields and display codes are given below.

Search Field	Search Code	Search Example	Display Code
Cited reference	/RE	S BYRON B, 1985/RE	RE
Cited reference author	/RAU	S CARTER A/RAU	RAU
Cited reference patent	/RPN	S US556669/RPN	RPN
Cited reference publication volume	/RVL	S 235/RVL	RVL
Cited reference publication year	/RPY	S RPY=1996	RPY
Cited reference work	/RWK	S Nature/RWK	RWK

The default display format, DISPLAY L(*ist number*), in SCISEARCH provides the accession number (AN), genuine article number (GA), title (TI), author (AU), corporate source (CS), country name of author (CYA), source (SO), document type (DT), file segment (FS), language

(LA), and cited reference count (REC). The DISPLAY ALL command also gives the abstract (AB), classification code (CC), supplementary term or author key word (ST), supplementary term plus or Key Words Plus (STP), research front (RF), table of cited references (RE). Any of these individual data fields can be accessed with the DISPLAY command.

(Copyright 1998 by the American Chemical Society and reprinted with permission.)

INTERNET

STN SCISEARCH Database Summary Sheets can be accessed at:

http://info.cas.org/ONLINE/DBSS/scisearchss.html

Dialog Information Retrieval Service Blue Sheets can be accessed at:

http://library.dialog.com/bluesheets

The Institute for Scientific Information maintains a web site at: http://www.isinet.com. Information is available on **Science Citation Index**, Web of Science (internet access to databases), the institute and its products.

CD-ROM

The *Science Citation Index* CD-ROM Edition and *Science Citation Index* CD-ROM with Abstracts Edition contain all the information available in the print edition. They can be searched by title, cited reference, author, key word, journal, and related citations.

REFERENCES

Haylock, John Richard, "Guide to Science Citation Index," Sunderland Polytechnic Library, 1974 (Z7401.H38 or Q158.5).

ORGANIZATION

The **Science Citation Index** is published bimonthly with annual and five-year cumulations beginning in 1965. Two ten-year cumulations, 1945–1954 and 1955–1964, are also available.

EXAMPLE ENTRY

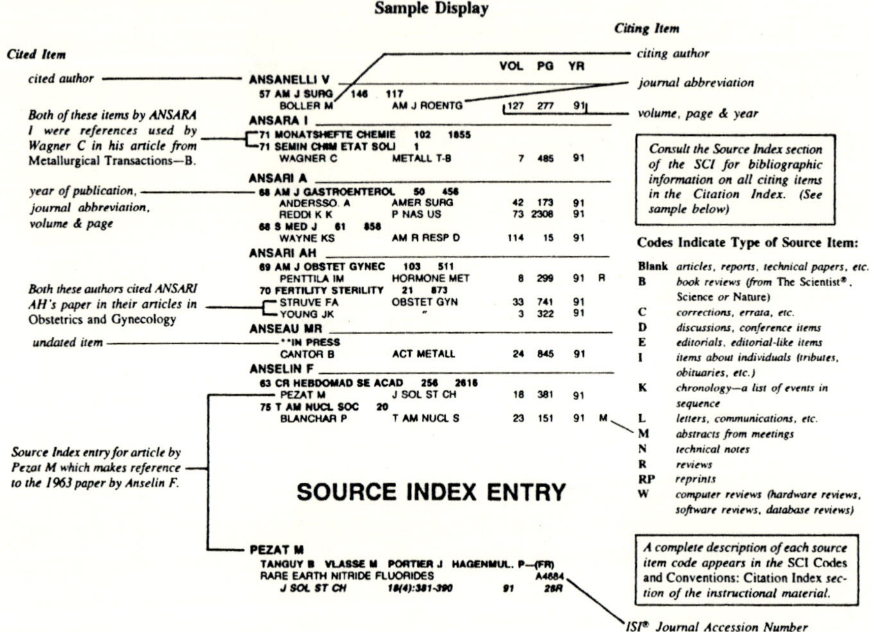

B.33.
Scientific Dictionaries

Books

Chemist's Dictionary. New York: Van Nostrand.

Ashford, Robert D. *Ashford's Dictionary of Industrial Chemicals: Properties, Production, Uses,* London: Wavelength Publications, 1994.

Bennett, H. Harry. *Concise Chemical and Technical Dictionary,* 4th enl. ed. New York: Chemical Publishing Co., 1986 (QD5.B4 1986).

Crouse, W. H. *McGraw-Hill Encyclopedia of Science and Technology,* 1987 (Q121 M3).

Grant, Julius. *Chemical Dictionary.* New York: McGraw-Hill.

Hawley, Gessner G. *The Condensed Chemical Dictionary.* New York: Van Nostrand Reinhold Co., 1981 (QD5.C5 1981).

Howard, Philip H., Neal, Michael. *Dictionary of Chemical Names and Synonyms.* Chelsea, Mich.: Lewis Publishers, 1992 (TP9.H65 1992).

Lewis, Richard J. Sr. *Hawley's Condensed Chemical Dictionary.* New York: Van Nostrand Reinhold, 1993 (QD5.C5 1993).

Morris, Christopher. *Academic Press Dictionary of Science and Technology.* San Diego, Calif.: Academic Press, 1992 (Q123.A33 1992).

Rose, Arthur and Elizabeth. *The Condensed Chemical Dictionary.* New York: Reinhold Publishing Corp., 1956 (QD5.C5 1956).

Walker, Peter M. B. *Chamber's Science and Technology Dictionary.* Cambridge, New York: Chambers-Cambridge, 1988 (T9.C5 1988).

INTERNET RESOURCES

American English Dictionary

       gopher://odie.niaid.nihgov/77/deskref/.Dictionary/enquire

Encyclopedia Britannica

       http://www.eb.com:180/eb.html

Virtual Reference Desk

       http://thorplus.lib.purdue.edu/reference/index.html

Webster's Dictionary

http://c.gp.cs.cmu.edu:5103/prog/webster?

http://www.m-w.com/dictionary.htm

One Look Dictionary

http://www.onelook.com

B.34.

Theilheimer's Synthetic Methods of Organic Chemistry:

QD262.T313

- **new methods**
- **significant variations on known methods**
- **reviews**

B.34.a.   OVERVIEW

Information on the preparation of reagents and special apparatus for conducting reactions is also included in the series.

EASY ACCESS

— For a chemical transformation, search the Subject Index for the product, checking both the *from* subheading and the *startg. m. f.*(starting material for) subsections.
— Reviews can be found in the Subject Index under the functional group or transformation of interest.
— Theilheimer's organizational symbols are explained and indexed in Sections B.34.b. and B.34.c.

INDEXING

Cumulative Indexes for each five years of the series are included in each fifth volume (i.e., Volumes 5, 10, 15 . . .). The subject indexes are also cumulated in two-year intervals in every second and fourth volume between the five-year cumulations. That is, all volumes numbers ending in 2, 4, 7, or 9 contain two-year cumulative subject indexes.

The **Formula Index of Functional Combinations** lists the molecular formula with the names of the functional groups whose synthesis has been described in the volume. The elements, are listed in alphabetical order, except for carbon which is listed last. Hal is used for all halogen elements, and hydrogens and alkyl groups are ignored. For example: an unsaturated carbonyl $\underline{O}=\underline{C}-\underline{C}=\underline{C}$ has a formula of OC_3.

The **Reagents Index** lists all the chemical reagents described in the series volume. The reagents are ordered according to the periodic system by the most significant element used in the chemical reaction (i.e., $Na_2Cr_2O_7$ under Cr).

The **Subject Index** lists compounds, reagents, and methods with emphasis placed on the functional group that is changed during a reaction. The subjects are subdivided into:

derivs — derivatives;

from — methods of preparing a particular substance from a starting functional group;

startg. m. f — the use of a starting material for the preparation of another substance.

Entries appear either as: Volume (in bold type), abstract number; or Volume (in bold print), abstract number, s (Supplementary Reference in) Volume (in bold print).

The **Supplementary References Index** serves to update literature reported in earlier volumes. The index is divided into volume numbers with columns listing the abstract numbers correlated to the later volume (in italic) and PAGE where the method is revisited.

The **Systematic Survey**, located in the back of each volume, is a table of contents of the reaction symbols reported.

COMPUTER SEARCH

Theilheimer database or *Derwent Journal of Synthetic Methods* File is available through STN International by typing FILE DJSMDS (subscriber file) or FILE DJSMONLINE (non-subscriber file). It is a key word and a structure-searchable chemical reactions database. The Basic Index (BI) contains *Derwent Journal of Synthetic Methods* Registry Numbers for all reaction participants, and single words from the abstract (AB), classification codes (CC), key words (KW), notes (NTE), and title (TI). Example search fields and display codes are given below.

Search Field Name	Search Code	Search Example	Display Code
Catalyst chemical			
name	/CAT.CN	S TICL3/CAT.CN	RX format
Key word	/KW	S PROTECTION/KW	KW
Product DJSM			
Registry No	/PRO	S 200/PRO	RX format
Product key word	/PRO.KW	S PURINE/PRO.KW	PRO.KW
Reactant key word	/RGT.KW	S STEROID/RCT.KW	RCT.KW
Solvent chemical name	/SOL.CN	S NH3/SOL.CN	RX format

The default display format, DISPLAY L(*ist number*), in DJSM provides information on the Journal, accession number (AU), title of document (TI), author names (AU), literature source (SO), volume/issue (VI), patent assignee (PA), patent number (PI), and other sources (OS). The DISPLAY ALL command also gives the abstract text (AB), classification code [CA section and cross-references] (CC), reaction symbol (SYM), key words (KW), notes (NTE), number of steps (NS), and the reaction map (RX). Any of these individual data fields can be accessed with the DISPLAY command.

(Copyright 1998 by the American Chemical Society and reprinted with permission.)

INTERNET

STN DJSMDS Database Summary Sheets can be accessed at:

http://info.cas.org/ONLINE/DBSS/djsmdsss.html

Derwent JSM Database information is available at:

http://library.dialog.com/bluesheets

LITERATURE DESCRIPTION

"Theilheimer's Synthetic Methods of Organic Chemistry, A Guide for Users" compiled by Alan F. Finch and Paul R. Mitchell. Basell: Karger: 1982.

ORGANIZATION

The Synthetic Methods are arranged in a reaction classification order by reaction symbols indicated at the top of each page. The first part of the symbols denotes the chemical bonds formed during the reaction with the elements listed as H, O, N, Hal(ogen), S, Rem(aining elements), and lastly C. The next symbol denotes the chemical method [addition ($\Downarrow$), rearrangement ($\cap$), exchange ($\downarrow\uparrow$), and elimination ($\Uparrow$)]. The final symbol denotes the bonds changed or eliminated. For example: the symbols HC$\Downarrow$OC stand for the formation of an H−C bond by addition to a O−C bond; that is, alcohol from reduction of an aldehyde. Each method is subdivided on the basis of reagents used. And each new abstract is given an abstract number. Numbers are assigned in sequential order for each volume.

EXAMPLE ENTRY

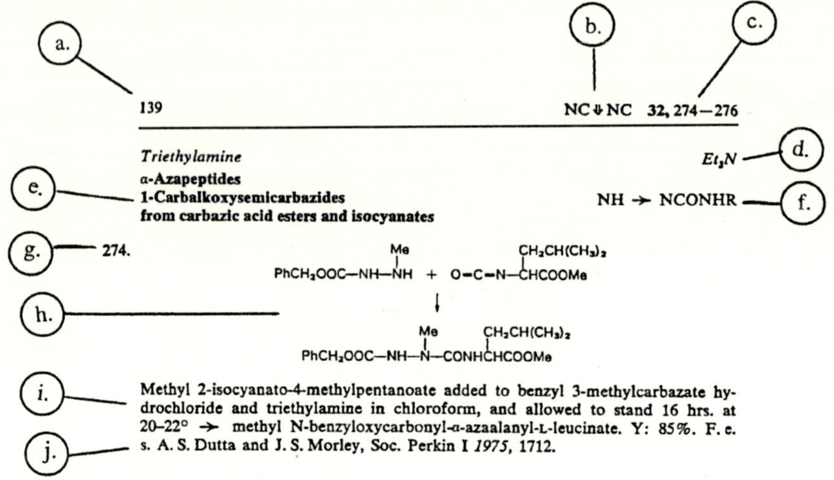

a.   page number
b.   Theilheimer Systematic Classification
c.   volume number, abstract numbers on this page
d.   auxiliary reagent
e.   reaction title
f.   general reaction
g.   abstract number
h.   reaction scheme
i.   experimental details
j.   citation

(Reprinted from *Theilheimer's Synthetic Methods of Organic Chemistry* with permission of S. Karger Publishers, Inc., Basel.)

B.34.b.   INDEX TO THEILHEIMER'S SYSTEMATIC CLASSIFICATION

A

acetic acid as reagent   NC⇑O   CC↓↑O   CC⇑O
acetic anydride as reagent   OC↑O   NC⇑O   CC⇑O   CC↓↑O
acetonylation   CC↓↑Hal
acetylene derivatives; hydrogenation   HC⇓CC
acid hydrolysis   HO↓↑C
acid to ethylene derivatives   CC⇑O
acid chlorides from acids   HalC↓↑O
acidolysis   OC↓↑S
acids to acid chlorides   HalC↓↑O
acids to aldehydes   HC⇑O
acids to hydrocarbons (decarboxylation)   HC⇑C
acids from acids   CC↓↑C
acylamines to amines   HN↓↑C
acylation of hydroxy compounds   OC↓↑Hal   OC↓↑O
acylation at oxygen   OC↓↑O
acylation at carbon   CC↓↑O
acylation at nitrogen   NC↓↑O HN↓↑C
O-acylation of hydroxy compounds   OC↓↑O   OC↓↑Hal
addition   ⇓
addition of halogens   HalC⇓
addition to double bond of hydroxyl groups   OC⇑CC
alcohols from esters, hydrolysis   HO↓↑C
alcohols from oxo compounts   HC⇓OC
alcohols to oxo compounts   OC⇑H
alcohols from halides   OC↓↑Hal
alcohols to hydrocarbons   HC⇑O
alcohols to alkyl halides   HalC↓↑
alcohols to aldehydes   OC⇑H (OC↓↑H)
alcohols from ketones   HC⇓OC
alcohols from esters via hydrolysis   HO↓↑C
alcohols to ketones   OC⇑H(OC↓↑H)
aldehydes from alcohols   OC⇑H(OC↓↑H)
aldehydes from acids   HC⇑O
aldehydes to acids by Reformatski reaction   CC↓↑Hal
aldehydes from hydrocarbons   OC↓↑H
Aldol condensation   CC⇑OC
Algar-Flynn oxidation   OC⇑H
alkyl halides to thioethers   SC↓↑Hal
alkyl halides from alcohols   HalC↓↑O

alkylation CC↓↑Hal
aluminium alcoholate HC⇓OC
aluminum chloride HO↓↑C CC↓↑O CC↓↑Hal CC∩OC
amines from halides NC↓↑Hal
amines from acyl amines HN↓↑C
amines from nitro compounds HN↓↑O
amines from nitriles HC⇓NC
amines to hydrocarbons HC↓↑N
amines to halogen compounds HalC↓↑N
Arbuzov-Michaelis rearrangement NRem↓↑C ORem↓↑N
Arndt-Eistert reaction CC↓↑Hal
Arndt-Eistert synthesis CC↓↑O
aromatization CC⇑H
avan Alphen rearrangement NC∩CC

B

Baeyer-Villiger oxidation OC⇓CC
Bamford-Stevens reaction CC⇑N
barium methylate HO↓↑C
Bates reagent OC⇑S
Beckmann fragmentation NC⇑O
Beckmann rearrangement NC⇓CC
Bender's salt SC↓↑Hal
Benkeser reduction HC⇓CC
Birch reduction HC⇓CC CC⇑C
Bischler-Napieralski ring closure CC⇑O
N-bromosuccinimide introduction of double bond CC⇑H
N-bromosuccinimide as reagent HalC↓↑H
boron trifluoride CC↓↑O
Bradsher ring closure CC↓↑Hal
Brown hydration OC⇓CC
Buck-Kobrich reaction OC↓↑Hal

C

C-chain extension CC⇓ CC↓↑
Cannizzaro reaction OC∩HC
carbalkoxylation NC↓↑Hal
carbohydrate deacylation HO↓↑C
carbonylation of alcohols CC⇓OC
carbonylation of amines NC↓↑O NC↓↑H
carbonylolefination CC↓↑C
chain extension CC⇓ CC↓↑
Claisen condensation CC⇓OC

Claisen rearrangement-amines   CC∩NC
Claisen rearrangement   CC∩OC
cleavage of ethers   HO↓↑C
cleavage   ⇑
Clemmensen reduction   CC⇓OC   HC↓↑O
condensation   CC↓↑O
Cope rearrangement   CC∩CC   CC∩OC
Cope elimination   CC⇑N
copper salts as additional reagents   HalC↓↑N   HC⇑C
copper cyanide   CC↓↑Hal
Cornforth oxazole rearrangement   NC∩OC
Cory-Winter elimination   CC⇑O
cuprous oxide   HC↓↑N
Curtius rearrangement   NC⇑S   NC⇓CC
cyclocarbonylation   CC⇓OC   CC⇓CC
cyclodehydrogenation   CC⇑H
cyclodehydrohalogenation   CC⇑Hal
cycloisomerization   CC∩HC

D
Dakin oxidation   OC⇓HC
Dakin-West reaction   CC↓↑C
Darzens reaction   CC↓↑Hal
Darzens-Claisen reaction   CC↓↑Hal   CC↓↑O
deacetylation   HO↓↑C
N-deacylation   HN↓↑C
N-deamination   HC↓↑N
O-debenzylation   CC↓↑O
decarboxamidation   HC⇑C
decarboxylation   HC⇑C
degradation by oxidation   OC↓↑C
degradation of side=chains   OC↓↑C
dehalogenation   CC⇑Hal
dehalogenation of 1,2-dihalides   CC⇑Hal
dehydration of hydroxy to oxo compounds   OC⇑H
dehydration   HC⇑O   CC⇑O   CC⇑H
dehydration of alcohols   CC⇑O
dehydrogenation   CC⇑H
dehydrohalogenation   CC⇑Hal   CC⇑H
desilylation   HC↓↑Rem   HO↓↑Rem
O-desilylation   CC∩OC
dethiolation   CC⇑S
diazoketones from acid chlorides   CC↓↑Hal

Dieckman cyclization   CC⇑O
Dieckmann-Komppa ring closure   CC↓↑O
diene synthesis   CC⇓CC
dimethyl sulfate   OC↓↑S
Dimroth rearrangement   NC∩NC

E
electrolytic reduction   HC↓↑O
elimination to unsaturated bonds   CC⇑
elimination from halides   ⇑Hal
Emde degradation   HC⇓NC
Emilewicz-Kostanecki ring closure   OC⇑Hal
English-Zimmerman fragmentation   OC∩CC
Eschenmoser fragmentation   CC⇑O
esterification of acids   OC↓↑O
esters to oxo compounds   CC⇑O
esters hydrolysis   HO↓↑C
esters from acids (esterification)   OC↓↑O
esters to ethylene derivatives   CC⇑O
esters to hydrocarbons   HC↓↑C
esters from acids   OC↓↑O
ether cleavage   HO↓↑C
ethers   OC↓↑Hal   OC↓↑N   OC↓↑S
ethylene derivatives via cleavage of halogen acids   CC⇑Hal
ethylene derivatives via cleavage of hydrogen   CC⇑H
ethylene derivatives to glycols   OC⇓CC
ethylene derivatives from 1,2-dihalides   CC⇑Hal
ethylene derivatives from alcohols   CC⇑O
ethylene to halides   HalC⇓CC
ethylene derivatives from esters   CC⇑O
ethylene to 1,2-dihalides   HalC⇓CC

F
Favorskii rearrangement   OC↓↑Hal
Fenton's reagent   CC↓↑C
Ferrario phenoxathin ring closure   SC↓↑H
Fisher indole ring closure   CC↓↑C   CC↓↑N   CC⇑N
Fisher indole ring synthesis   CC⇑N
fluorination   HalC↓↑H
Friedel-Crafts alkylation   CC↓↑Hal
Friedlander pyridine ring synthesis   CC↓↑O

G

Gabriel synthesis   NC↓↑Hal
Gabriel Colman rearrangement   CC∩NC
Gattermann formylation   CC↓↑H
Gattermann-Koch type isoquinoline synthesis   CC↓↑O
glycols from ethylene derivatives   OC⇊CC
glycoside cleavage   HO↓↑C
glycosylation   OC↓↑C
Gomberg-Bachman reaction   CC↓↑N
Grignard reaction   CC⇊OC   CC⇊NC   CC↓↑Hal
Grob fragmentation   CC⇑S   CC⇑Hal

H

halides from alcohols   HalC↓↑O
halides from amines   HalC↓↑N
halides from ethylene derivatives   HalC⇊CC
haloform reaction   HalC↓↑C
halogen addition to double bonds   HalC⇊CC
halogen addition   HalC⇑
halogen cleavage   CC⇑Hal
halogen elimination   ⇑Hal   CC⇑Hal
halogenation of double bonds   HalC⇊CC
halogenation   HalC↓↑H
halogenolysis   HalC↓↑C
halogens as additional reagents   SC↓↑H   OS⇊S
Hassner-Ritter reaction   NC⇊CC
Hayashi rearrangement   CC⇑O
Heck arylation   CC↓↑Hal   CC↓↑Rem
Helferich-Schmitz-Hillberecht reaction   OC↓↑C
Hilbert-Johnson reaction   NC↓↑C
Hoch-Campbell aziridine ring closure   CC↓↑Hal
Hoffmann degradation   CC⇑N
Hofmann-Loffler ring closure   NC⇑H
Horner synthesis   CC↓↑Rem
Hugershoft benzothiazole ring closure   SC⇑H
Hurtley arylation   CC↓↑Hal
hydration of acid chloride   HC↓↑Hal
hydrobenzene rings via diene synthesis   CC⇊CC
hydrocarbons to ketones via Friedel-Crafts   CC↓↑Hal
hydrocarbons from aldehydes   HC↓↑O
hydrocarbons to alcohols   OC↓↑H
hydrocarbons from alcohols   HC⇑O
hydrocarbons from amines   HC↓↑N

hydrocarbons from acids (decarboxylation)   HC⇑C
hydrocarbons from oxo compounds   HC↓↑O
hydrocarbons from hydroxy compounds   HC⇑O
hydrocarbons from ketones   HC↓↑O
hydrocarbons from esters   HC↓↑C
hydrogen fluoride in condensations   CC⇑O
hydrogen peroxide   OS⇓S   OC⇓CC
hydrogen halides cleavage from halogen compounds   CC⇑Hal
hydrogenation of CC double bonds   HC⇓CC
hydrogenation   HC⇓CC   HO   NH   HC
hydrogenolysis   HC↓↑O   HC↓↑N
hydrolysis of acyl amines   HN↓↑C
hydrolysis of acetals   HO↓↑C
hydrolysis of esters   HO↓↑C
hydrolysis of sulfonic acid esters   HO↓↑S
hydrolysis   HO↓↑C   HN↓↑C
hydrometallation   HC⇓CC
hydrosilylation   RemC⇓CC
hydroxy compounds to oxo compounds   OC⇑H
hydroxylation   OC⇓HC

K
Kakis reaction   CC↓↑O
keto acids to acids   CC⇑C
ketones from alcohols   OC⇑H   (OC↓↑H)
ketones to alcohols   HC⇓OC
Knoevenagel synthesis of unsaturated compounds   CC↓↑O
Knoevenagel condensation   CC↓↑O
Knoevenagel-Cope condensation   CC↓↑O
Knorr pyrrole synthesis   CC↓↑N
Kolbe electrolysis   CC↓↑C
Konigs-Knorr reaction   OC↓↑Hal
Kostanecki-Robinson reaction   CC↓↑O
Krohnke reaction   CC↓↑N

L
lactone annelation   CC⇑C
Lander rearrangement   NC∩OC
lithium   CC⇓OC

M
magnesium   OC↓↑Hal   CC↓↑Hal   CC⇓NC
Marschalk reaction   CC↓↑O

McMurry's reagent   CC↓↑O
Meerwein-Ponndorf reduction   HC⇓OC
mercaptans to thio ethers   SC↓↑Hal
mercury   OC⇓CC
Merwein, Ponndorf-Verley reduction   HC⇓OC
Meyer-Schuster rearrangement   CC∩HO
Michael addition   CC⇓CC
Moffatt oxidation   OC⇑H
Mohlau-Bischler indole ring closure   CC↓↑O
Mukaiyama reagent   OC↓↑O

N

Neber rearrangement   OC↓↑Hal
Neber rearrangement   NC↓↑O
Nef reaction   OC↓↑N
nickel   HH↓↑O; HC↓↑O; HC↓↑Hal; HC↓↑S; HC⇓OC; HC⇓CC
nitration   NC↓↑O   NC↓↑H
nitriles to amines   HC⇓NC
nitro compounds to amines   HN↓↑O
nucleus hydration   HC⇓CC

O

Oppenauer oxidation   HC⇓OC
Orthoester Claisen rearrangement   CC↓↑O
osmium tetroxide   OC⇓CC
oxidation of alcohols to ketones   OC⇑H
oxidation   OC↓↑H   OC⇑C   OC⇑H   HalC↓↑H   ON; OS; OC
oxidative degradation   OC↓↑C
oxo compounds from hydrocarbons   HC↓↑O
oxo from hydroxy   OC⇑H
Oxy-Cope rearrangement   CC∩HO
ozonolysis   OC↓↑C   OC∩CC

P

palladium catalyst   HC↓↑O   HC↓↑Hal   CC⇑H
  HN↓↑C   HC⇓OC
palladium on carbon   HC⇓OC   HC⇓CC   HC↓↑Hal
palladium-barium sulfate   HC↓↑O   HC↓↑Hal
permanganate   OC↓↑H   OS⇓S
Peterno-Buchi reaction   OC∩CC
phenol from phenol ethers   HO↓↑C

phenol ethers cleavage   HO↓↑C
phosphorous oxide chloride   HalC↓↑O
phosphorus halides   HalC↓↑O   CC⇑O
phosphorylation   ORem↓↑Hal   ORem↓↑N   ORem↓↑O
Pictet-Gams isoquinoline ring closure   CC⇑O
Pictet-Springer ring closure   CC↓↑O
platinum   HC⇓OC   HC⇓CC   HC↓↑O   HC⇓NC
polyphosphoric acid   CC⇑O
Pomeranz-Fritsch isoquinoline ring closure   CC⇑O
Prevost reagent   OC⇓CC   OC↓↑CC
Prins reaction   CC⇓CC
Pummer rearrangement   OC⇑O
pyridine hydrochloride as additional reagent   HO↓↑C

Q
Quasi-Favorski rearrangement   NC↓↑Hal
N-Quaternization   NC↓↑O   Het⇓N

R
radical halogenation   Hal↓↑H
reduction with LiAlH$_4$ HC⇓OC   HC⇓NC   HC⇓CC   HC↓↑O
reduction with NaBH$_4$ HC⇓OC   HC⇓NC   HC↓↑O
reduction with Raney Nickel   NH↓↑O HC⇓OC HC⇓CC
    HC↓↑O HC↓↑S
Reformatski synthesis   CC↓↑Hal   CC↓OC·Zn
Reissert reaction   RemC↓↑C
replacement of hydroxyl by hydrogen   HC⇑O
replacement of halogen by thioether group   SC↓↑Hal
replacement of halogen by alkoxyl group   OC↓↑Hal
replacement of hydrogen by halogen   HalC↓↑H
replacement of halogen by hydrogen   HC↓↑Hal
replacement of hydroxyl by halogen   HalC↓↑O
replacement of sulfur by hydrogen   HC↓↑S
replacement of amino group by halogen   HalC↓↑N
replacement of halogen by carboxyl group   OC↓↑Hal
replacement of amino group by hydrogen   HC↓↑N
replacement of halogen by amino group   NC↓↑Hal
replacement of hydroxyl by amino   NC↓↑O
replacement of hydrogen by hydroxyl   OC⇓HC
retro-Claisen rearrangement   OC∩CC
retro-Dimroth rearrangement   NC∩ON
retro-Mannich reaction   HC↓↑C

retro-Michael conversion CC↓↑C
retrocarbene reaction CC⇈C
retrodiene scission CC⇈C
Ritter reaction NC⇊NC
Robinson annealation CC↓↑O
Rosenmund-Zetzsche reduction HC↓↑Hal

S
Sandmeyer reaction HalC↓↑N
Scherer ring contraction OC↓↑Hal
selenium in dehydration CC⇈H
selenium dioxide OC↓↑H
Semmler-Wolff reaction CC⇈N
Shapiro reaction CC↓↑N
side-chain oxidation OC↓↑C
O-Silylation ORem↓↑O
silver oxide OC↓↑N
silver carbonate OC↓↑Hal
Silyl-Pummerer rearrangement OC∩RemC　　OC⇈Hal
Simmons-Smith cyclopropane ring synthesis CC↓↑Hal
Skraup synthesis CC↓↑O
Smiles rearrangement NC∩OC
sodium sulfide HN↓↑O
sodium amide in condensations CC↓↑Hal
sodium in reductions HC↓↑O　　HC⇊NC
sodium in condensations CC↓↑O
sodium nitrite OC↓↑N; ON↓↑H; NN↓↑O;
　NC↓↑O; CC↓↑N; CC∩CC
sodium hyposulfite HN↓↑O
sodium hypochlorite in oxidation OC↓↑C
Sommelet-Hauser rearrangement CC∩NS　　CC∩NC　　CC↓↑C
stannic chloride CC⇈Hal
Stobbe condensation CC↓↑O
Strecker reaction CC↓↑O
sulfur in dehydration CC⇈H
sulfuric acid in condensations CC↓↑O
synthesis CC⇊　　CC↓↑

T
thiazole ring closure SC↓↑Hal
thio ethers from mercaptans SC↓↑Hal
thio ethers from alkyl halides SC↓↑Hal

thionyl chloride   HalC↓↑O   Hal↓↑O
p-toluensulfonic acid as a reagent   OC↓↑O
transacetallation   OC↓↑C
transamidation   HN↓↑C   NC↓↑C   NC∩HN
transesterification   OC↓↑C
transhalogenation   HC↓↑Hal
translactonization   HC⇓OC   OC∩HO
transacetallation   OC⇓CC   CC↓↑Rem
Treibs reaction   OC⇓HC

U
Ugi reaction   CC↓↑O

V

Vilsmeier aldehyde synthesis   CC↓↑N

W

water addition   OC⇓
Wichterle ring closure   CC⇑Hal
Willgerodt-Kindler reaction   NC↓↑O   OC∩HO
Wittig synthesis   CC↓↑Rem
Wolff rearrangement   OC↓↑N
Wolff-Kishner   HC↓↑O
Wolff-Kishner reduction   HC↓↑O

Y
Yates-Hyre synthesis   CC↓↑C

Z
Ziegler ring closure   CC∩NC
zinc amalgam   HC↓↑O
zinc salts   CC↓↑N⇑   CC↓↑O
zinc   CC⇑Hal; HN⇓NN; HN↓↑O; HC↓↑O; CC↓↑Hal
zinc chloride   HalC↓↑O
Zip reaction   NC∩HN

B.34.c. THEILHERMER'S SYMBOL — FUNCTIONAL GROUP TRANSFORMATIONS

FROM ↓ \ TO→	.Alcohols	Aldehydes	Amines	Esters	Ethers	Halides	Hydrocarbons	Ketones	Olefins
Acids		HC⇑O		OC↓↑O			HC⇑C		CC⇑O
Alcohols		OC⇑H	NC↓↑O		HO↓↑C	HalC↓↑O	HC⇑O	OC⇑H	CC⇑O
Aldehydes	HC⇓OC						HC↓↑O		
Amines					OC↓↑N	HalC↓↑N	HC↓↑N		
Esters	HO↓↑C	CC⇑O					HC↓↑C	CC⇑O	CC⇑O
Halides	OC↓↑Hal	OC↓↑Hal	NC↓↑Hal		OC↓↑Hal		HC↓↑Hal	OC↓↑Hal	CC⇑Hal
Hydrocarbons	OC↓↑H	OC↓↑H				HalC↓↑H		OC↓↑H	CC⇑H
Ketones	HC⇓OC						HC↓↑O		
Olefins	OC⇓CC					HalC⇓CC			

B.35.
Translations

Books

FRENCH

DeVries, L. *French-English Dictionary.* New York: McGraw-Hill, 1962.

Patterson, A. M. *German-French Dictionary for Chemists*, 2nd ed. New York: Wiley, 1954 (QD5.P25 1954).

GERMAN

Condoyannis, G. *Patterson's German-English Dictionary for Chemists.* New York: Wiley, 1991 (QD5.P3 1991).

DeVries, L. *German-English Science Dictionary.* McGraw-Hill, New York:

Neville, H. H., Johnston, N. C., Boyd, G. V. *A New German/English Dictionary for Chemists.* London: Blackie, 1964 (QD5.N47).

Patterson, A. M. *German-Englilsh Dictionary for Chemists*, 3rd ed. New York: Wiley, 1950 (QD5.P3 1950).

Webel, A. *A German-English Technical and Scientific Dictionary.* New York: Dutton, 1952 (Q123.W3 1952).

JAPANESE

Tung, L. W. *Japanese/English English/Japanese Glossary of Scientific and Technical Terms.* New York: Wiley, 1995.

RUSSIAN

Blum, Alexander. *Concise Russian-English Scientific Dictionary for Students and Research Worker.* Oxford, New York: Pergamon Press, 1965 (Q123.B55).

Callaham, L. I. *Russian-English Chemical and Polytechnical Dictionary.* New York: Wiley, 1975.

Hoseh, Mordecai, Hoseh, Melanie L. *Russian-English Dictionary of Chemistry and Chemical Technology.* New York: Reinhold Publishing Corp., 1964 (QD5.H6).

Katzner, K. *English-Russian/Russian-English Dictionary*. New York: Wiley, 1994 (PG2640.K34 1994).

MULTILANGUAGE

Dorian, A. F. *Six-Language Dictionary of Electronics, Automation and Scientific Instruments: A Comprehensive Dictionary in English, French, German, Italian, Spanish and Russian*. London: Iliffe Books; Englewood Cliffs, N.J.: Prentice-Hall, 1962 (TK7804.S5).

Internet Resources

FRENCH-ENGLISH DICTIONARIES

French Language Resources

http://www.mlab-power3.viah.fa/EnglishFrench/ef.html

ARTFL Project: French-English Dictionary Form

http://www.humanities.uchicago.edu/forms unrest/FR-ENG.html

French-English Dictionary

http://www.lib.ua.edu/dict4.htm

GERMAN-ENGLISH DICTIONARIES

LEO German-English Dictionary

http://www.leo.org/cgi-bin/dict-search

German-English On-line Dictionary

http://www.travlang.com/GermanEnglish

IEE German-English

http://www.iee.et.tu-dresden.de/cgi-bin/cgiwrap
/wernerr/search.sh

JAPANESE-ENGLISH DICTIONARIES

Japanese-English Dictionary

http://www..wg.omron.co.jp/cgi-bin/j-e

RUSSIAN-ENGLISH DICTIONARIES

Russian-English Dictionary

http://solar.rtd.utk.edu/cgi-bin/mtrans

Russian-English Dictionary

http://www.elvis.ru/cgi-bin/mtrans

MULTILANGUAGE DICTIONARIES

A Web of On-line Dictionaries

http://www.bucknell.edu/~rbeard/diction.html

Yahoo's List of Dictionaries

http://www.yahoo.com/Reference/Dictionaries

List of Dictionaries

http://math-www.uni-paderborn.de/HTML/Dictionaries.html

THOR+: The Virtual Reference Desk — Dictionaries

http://thorplus.lib.purdue.edu/reference/dict.html

Dictionaries — Education

http://www.virstadskola.pp.se/lexicon.html

Research It!

http://www.itolls.com/research-it

C.1.

Organic Chemistry Reviews

The following sections may be helpful in finding Organic Chemistry
Review Articles
> Annual Reports in Organic Synthesis
> Compendium
> Comprehensive Organic Chemistry
> Encyclopedia of Reagents for Organic Synthesis
> Larock's Comprehensive Organic Transformations
> Organic Reactions
> Science Citation Index
> Theilheimer

Books

East, Michael B., Ager, David J. *Desk Reference for Organic Chemists.*
Malabar, Fla.: Krieger Publishing Company, 1995.

Hassner, Alfred, Stumer, C. *Organic Syntheses Based on Name Reactions and Unnamed Reactions.* Oxford, UK; Tarrytown, N.Y.: Pergamon, 1994 (QD262.H324).

Krauch, Helmut, Kunz, Werner. *Organic Name Reactions.* New York: Wiley, 1964 (QD291.K713 1964).

Mundy, Bradford P., Ellerd, Michael G. *Name Reactions and Reagents in Organic Synthesis.* New York: Wiley, 1988 (QD291.M86).

Surrey, Alexander Robert. *Name Reactions in Organic Chemistry.* New York: Academic Press, 1961 (QD291.S87 1961).

Internet Resources

Synthesis Reviews sponsored by Georg Thieme Verlag at

> http://www.chem.gal.ac.uk/~krisj/srev.htm

ChemKey "Review" Database for Mac can be downloaded at

> http://euch6f.chem.emory.edu/

Reaction Index, Name Reactions in Organic Chemistry

> http://www.pmf.ukim.edu.mk/PMF/Chemistry/reactions/rindex.htm

INDEX TO REVIEWS IN TETRAHEDRON AND SYNTHESIS (1972–1997)

A

acetals — *Synthesis* (1985), 592
acetals, coupling — *Synthesis* (1987), 1043
acetals, dithio α-oxoketene — *Tetrahedron* (1986), **42**, 3029
acetals, formamide — *Tetrahedron* (1979), **35**, 1675
acetals, lactam — *Tetrahedron* (1988), **44**, 5975
acetals, mono- — *Synthesis* (1985), 592
acetals, α-oxoketene — *Tetrahedron* (1990), **46**, 5423
acetals, synthesis from epoxides — *Synthesis* (1981), 501
acetogenins, natural source — *Tetrahedron* (1995), **51**, 4571
acetoxy ethers — *Synthesis* (1988), 95
acetylene equivalents — *Tetrahedron* (1984), **40**, 2585
N-acetylneuraminic acid — *Synthesis* (1991), 583
acids from carbonylation reactions — *Synthesis* (1973), 509
aconite alkaloids — *Tetrahedron* (1985), **41**, 485
acroleins, alkoxy — *Synthesis* (1987), 1
acyclic stereocontrol — *Synthesis* (1991), 594
acyl anions, from cyanohydrins — *Tetrahedron* (1983), **39**, 3207
acyl chlorides — *Synthesis* (1973), 189
acyl cyclopentenes, isomerization — *Tetrahedron* (1976), **32**, 641
N-acyliminium intermediates — *Tetrahedron* (1985), **41**, 4367
acyl transfer, enzyme-catalyzed — *Synthesis* (1992), 895
acylation — *Tetrahedron* (1976), **32**, 1943
acylations, Friedel-Crafts — *Synthesis* (1972), 533
N-acyliminium intermediates — *Tetrahedron* (1985), **41**, 4367
acylisocyanates, cycloadditions — *Synthesis* (1974), 461
acyloxylation at carbon — *Synthesis* (1973), 567
acyloxylation, substitutive at carbon — *Synthesis* (1972), 1
adamantane homologues — *Tetrahedron* (1980), **36**, 971
addition reactions, base catalyzed — *Synthesis* (1974), 309
addition reactions of silanes — *Synthesis* (1988), 263
addition reactions, tetracyanoethylene — *Synthesis* (1987), 749
addition reactions with ketenes — *Tetrahedron* (1986), **42**, 2587
addition reactions, with TCNE — *Synthesis* (1987), 749
additions, organocopper conjugate — *Synthesis* (1985), 364
adrenoceptor agonists — *Tetrahedron* (1991), **47**, 9953
alcohol, manipulation — *Tetrahedron* (1985), **41**, 643
alcohols, deoxygenation by radicals — *Tetrahedron* (1983), **39**, 2609
alcohols, deuterated — *Synthesis* (1972), 254

alcohols, oxidation   *Synthesis* (1990), 857
alcohols, oxidation   *Synthesis* (1997), 1115
alcohols, protection   *Tetrahedron* (1997), **53**, 13509
aldehydes, addition to   *Synthesis* (1992), 248
aldehydes, cheleselective reactions   *Synthesis* (1995), 745
aldehydes, reactions   *Tetrahedron* (1988), **44**, 4653
aldehydes, thioalkylation   *Synthesis* (1987), 589
3-alkoxy acroleins   *Synthesis* (1987), 1
alkali metal derivatives, poly   *Synthesis* (1977), 509
alkaloid biosynthesis   *Tetrahedron* (1991), **47**, 5945
alkaloids, aconite synthesis   *Tetrahedron* (1985), **41**, 485
alkaloids, anaatoxin-α   *Tetrahedron* (1996), **52**, 6025
alkaloids, aporphinoid   *Tetrahedron* (1984), **40**, 4795
alkaloids, *bis*-indole   *Tetrahedron* (1982), **38**, 223
alkaloids, camptothecin   *Tetrahedron* (1981), **37**, 1047
alkaloids, ellipticine   *Tetrahedron* (1986), **42**, 2389
alkaloids, ergot   *Tetrahedron* (1976), **32**, 873
alkaloids, ergot   *Tetrahedron* (1980), **36**, 3123
alkaloids, phenolic oxidation   *Synthesis* (1972), 657
alkaloids, polyhydroxylated   *Synthesis* (1995), 607
alkaloids, sesquiterpenoids   *Tetrahedron* (1996), **52**, 7087
alkenes   *Synthesis* (1996), 1
alkenes, amination   *Tetrahedron* (1983), **39**, 703
alkenes, aromatic substitution   *Synthesis* (1973), 524
alkenes, arylation, Pd-cat   *Tetrahedron* (1997), **53**, 7371
alkenes, bridgehead   *Tetrahedron* (1980), **36**, 1683
alkenes, Prins reaction   *Synthesis* (1977), 661
alkenes, trophiles   *Tetrahedron* (1996), **52**, 8001
alkenes, vinylation, Pd-cat   *Tetrahedron* (1997), **53**, 7371
alkenylation of amides and imides   *Synthesis* (1975), 685
alkoxyacroleins   *Synthesis* (1987), 1
alkoxyallenes   *Synthesis* (1993), 165
alkyl anion synthons   *Synthesis* (1988), 833
alkyl-radical reactions   *Tetrahedron* (1993), **49**, 1151
alkylation, intramolecular, spiro   *Synthesis* (1974), 383
alkylation of ketones   *Synthesis* (1983), 517
alkylation, α-thio, with aldehydes   *Synthesis* (1987), 589
alkylation, α-ureido, cyclic ureas   *Synthesis* (1973), 243
alkylation, with isoureas   *Synthesis* (1979), 561
alkylidenephosphoranes   *Synthesis* (1974), 775
alkynes, semihydrogenation   *Synthesis* (1973), 457
alkynes, cyclo-, synthesis   *Synthesis* (1972), 599

alkynes, metallation — *Synthesis* (1981), 841

allenes — *Tetrahedron* (1984), **40**, 2805

allenes, chiral — *Synthesis* (1973), 25

allenes, ketonic — *Tetrahedron* (1980), **36**, 331

allene oxides — *Tetrahedron* (1980), **36**, 2269

allenyl enolates — *Tetrahedron* (1997), **53**, 10197

allyl type organometallics, polar — *Tetrahedron* (1993), **49**, 10175

allylic alcohols, asymmetric epoxidation — *Synthesis* (1986), 89

allylic compounds, hydrogenolysis — *Synthesis* (1996), 1

allylic compounds, displacement — *Tetrahedron* (1980), **36**, 1

allylic silanes — *Synthesis* (1988), 263

allylic tin compounds — *Tetrahedron* (1993), **49**, 7395

allylpalladium compounds — *Tetrahedron* (1986), **42**, 4361

allylsilanes — *Synthesis* (1990), 969

allylsilanes — *Synthesis* (1990), 1101

alumina — *Tetrahedron* (1997), **53**, 7999

aluminium compounds — *Tetrahedron* (1988), **44**, 5001

aluminum compounds, reduction — *Synthesis* (1972), 217

Amanita muscaria pigments — *Tetrahedron* (1980), **36**, 2843

α-amino acids — *Tetrahedron* (1991), **47**, 6079

3-amino-2-azetidinones — *Tetrahedron* (1991), **47**, 7503

2-aminobenzimidazoles — *Synthesis* (1983), 861

amides, alkenylation — *Synthesis* (1975), 685

amides, hydrolysis — *Tetrahedron* (1975), **31**, 2463

amidines, bicyclic — *Synthesis* (1972), 591

amidyl radicals, reactivity — *Tetrahedron* (1978), **34**, 3241

amination, electrophilic — *Synthesis* (1991), 327

amination, mesitylenesulfonyl hydroxylamine — *Synthesis* (1977), 1

amine N-oxides — *Synthesis* (1993), 263

amines, allylic — *Synthesis* (1983), 685

amines, dealkylation — *Synthesis* (1989), 1

amines, from alkenes — *Tetrahedron* (1983), **39**, 703

amines, oxidation — *Synthesis* (1997), 1115

amines, tertiary, catalysts — *Tetrahedron* (1996), **52**, 8001

amino acids — *Tetrahedron* (1988), **44**, 5515

amino acids, azabicycloalkane — *Tetrahedron* (1997), **53**, 1 2798

amino acids, chiral, synthesis — *Synthesis* (1995), 607

amino acids, iron chelators — *Tetrahedron* (1995), **51**, 3939

amino acids, cyclopropane — *Tetrahedron* (1990), **46**, 2231

amino acids, didehydro — *Synthesis* (1988), 159

amino acids, hydroxy   *Tetrahedron* (1996), **52**, 7063
amino aldehydes, heteroannelation   *Tetrahedron* (1980), **36**, 2359
amino groups, replacement   *Tetrahedron* (1980), **36**, 679
amino alcohols, chiral synthesis   *Synthesis* (1982), 605
amino acids   *Tetrahedron* (1994), **50**, 1539
amino-oximes, α-hydroxyl   *Synthesis* (1986), 704
amino-phosphine derivatives   *Synthesis* (1986), 793
amino-*tert*-alkanes   *Synthesis* (1995), 1053
aminobenzophenones   *Synthesis* (1980), 677
aminobenzothiazoles   *Synthesis* (1983), 861
ammonium compounds, quaternary   *Synthesis* (1973), 441
ammonium formate   *Synthesis* (1988), 91
annulation, Robinson   *Synthesis* (1976), 777
anion synthons, alkyl   *Synthesis* (1988), 833
anions, acyl equivalents   *Tetrahedron* (1983), **39**, 3207
anions, di-   *Synthesis* (1982), 521
anions, di-, polyconjugated   *Tetrahedron* (1988), **44**, 6957
anions, vinyl   *Tetrahedron* (1988), **44**, 4653
anisatin   *Synthesis* (1995), 729
annulation   *Tetrahedron* (1976), **32**, 3
anodic substitution reactions   *Tetrahedron* (1976), **32**, 2185
anomeric effect, reverse   *Tetrahedron* (1995), **51**, 11901
anthocyanins and flavylium salts   *Tetrahedron* (1983), **39**, 3005
anthracene derivatives   *Tetrahedron* (1992), **48**, 3251
anthracyclinones   *Tetrahedron* (1990), **46**, 291
anti-AIDS drugs   *Synthesis* (1995), 1465
anti-HIV drugs   *Synthesis* (1995), 1465
antibiotics, carbapenum   *Tetrahedron* (1996), **52**, 331
antibiotics, hybrid   *Tetrahedron* (1991), **47**, 6045
antibiotics, β-lactam biosynthesis   *Tetrahedron* (1977), **33**, 1545
antibiotics, macrolide   *Tetrahedron* (1977), **33**, 3041
antibiotics, macrolide   *Tetrahedron* (1977), **33**, 683
antibiotics, macrolide   *Tetrahedron* (1985), **41**, 3569
antibiotics, orthosomycins   *Tetrahedron* (1979), **35**, 1207
anticancer drugs   *Synthesis* (1995), 1465
antitumor, antibiotics, enediyne   *Tetrahedron* (1996), **52**, 6453
arene chromium tricarbonyl   *Synthesis* (1993), 643
arenesulfonyloxy groups   *Tetrahedron* (1991), **47**, 1109
aromatic compounds, biosynthesis   *Tetrahedron* (1978), **34**, 3353
aromatic compounds, iodo-,   *Synthesis* (1988), 923
aromatic compounds, protective   *Synthesis* (1979), 921
  groups
aromatic compounds, poly   *Tetrahedron* (1988), **44**, 2093

aromatic hydrocarbons, polycyclic, oxidized metabolites — *Synthesis* (1986), 605

aromatic, photochemical hydroxylation — *Synthesis* (1974), 173

aromatic, polyfluoro-, — *Synthesis* (1976), 652

aromatic reactivity — *Tetrahedron* (1993), **49**, 4477

aromatic substitution, with oxazolines — *Tetrahedron* (1985), **41**, 837

aromatic substitution — *Tetrahedron* (1978), **34**, 2057

aromatic substitution by Pd salts — *Synthesis* (1973), 524

aromaticity — *Tetrahedron* (1988), **44**, 7427

aromatics by organoiron compounds — *Tetrahedron* (1983), **39**, 4027

arsanes, amino — *Synthesis* (1982), 173

arsenides, synthesis and reactions — *Synthesis* (1974), 328

artificial membranes, photosensitive — *Tetrahedron* (1994), **50**, 4039

aryl-aryl bond formation — *Tetrahedron* (1980), **36**, 3327

arylation — *Tetrahedron* (1988), **44**, 3039

C-aryl glycosides, synthesis — *Synthesis* (1994), 1

aryl halides — *Tetrahedron* (1984), **40**, 1433

arylation — *Tetrahedron* (1988), **44**, 3039

2-aryloxazolones — *Synthesis* (1975), 749

2-arylpropionic acids, — *Tetrahedron* (1986), **42**, 4095

aryls, bi-, Ullmann synthesis — *Synthesis* (1974), 9

arynes, hetero — *Tetrahedron* (1982), **38**, 427

asymmetric epoxidation of allylic alcohols — *Synthesis* (1986), 89

asymmetric hydrogenation — *Synthesis* (1981), 85

asymmetric reactions — *Synthesis* (1991), 665

asymmetric synthesis — *Synthesis* (1978), 329

asymmetric synthesis — *Tetrahedron* (1979), **35**, 2797

asymmetric synthesis — *Tetrahedron* (1980), **36**, 2

asymmetric synthesis — *Tetrahedron* (1986), **42**, 5157

asymmetric synthesis — *Synthesis* (1990), 541

asymmetric synthesis — *Tetrahedron* (1991), **47**, 6079

asymmetric synthesis, catalytic — *Synthesis* (1991), 1

asymmetric synthesis, industrial — *Tetrahedron* (1994), **50**, 3639

asymmetric synthesis, of lignans — *Tetrahedron* (1990), **46**, 5029

asymmetric synthesis, with camphor — *Tetrahedron* (1987), **43**, 1969

asymmetric synthesis with chiral boranes — *Tetrahedron* (1981), **37**, 3547

asymmetric synthesis, with pig liver esterases — *Tetrahedron* (1990), **46**, 6587

asymmetric thermal reactions — *Tetrahedron* (1993), **49**, 293

aza-crown compounds, aliphatic   *Synthesis* (1982), 997
aza-propeniminium units   *Synthesis* (1988), 655
2-azapropenes, 1,3,3-trichloro-   *Synthesis* (1972), 599
azadienes, Diels-Alder reactions   *Tetrahedron* (1983), **39**, 2869
azapeptides   *Synthesis* (1989), 405
azaphospholes, synthesis   *Synthesis* (1995), 361
azasugars, synthesis   *Synthesis* (1995), 607
azetidine ring   *Synthesis* (1995), 729
azide method in peptide synthesis   *Synthesis* (1974), 549
azides, lead IV, acetate   *Synthesis* (1972), 285
azido-tetrazolos, isomerization   *Synthesis* (1973), 123
azines   *Tetrahedron* (1988), **44**, 1
azines, reaction with dienophiles   *Synthesis* (1976), 349
azines, ring degeneration   *Tetrahedron* (1985), **41**, 237
1-azirines, cycloadditions   *Synthesis* (1975), 483
azobenzene, for artificial membranes *Tetrahedron* (1994), **50**, 4039
azodicarboxylate   *Synthesis* (1981), 1
azomethine   *Tetrahedron* (1976), **32**, 2165
azomethine imines   *Synthesis* (1973), 469
azomethines, with dienophiles   *Synthesis* (1976), 349
azomethine ylids   *Synthesis* (1973), 469

B
Barbier reaction, organomagnesium   *Synthesis* (1977), 18
Barbier-Grignard reactions   *Tetrahedron* (1996), **52**, 5643
base catalyzed addition reactions   *Synthesis* (1974), 309
bases, Mannich   *Synthesis* (1973), 703
Baylis-Hillman reaction   *Tetrahedron* (1996), **52**, 8001
benzene, photochemistry   *Tetrahedron* (1976), **32**, 1309
benzene, photochemistry   *Tetrahedron* (1977), **33**, 2459
benzenoids   *Tetrahedron* (1993), **49**, 9207
benzocyclobutenone   *Synthesis* (1978), 869
benzocyclopropyenyl cation and   *Tetrahedron* (1988), **44**, 1305
    anion
benzofuroxan   *Synthesis* (1975), 415
benzothiazole   *Synthesis* (1985), 586
benzotriazole   *Synthesis* (1994), 445
benzotriazole   *Tetrahedron* (1991), **47**, 2683
benzotriazole alkylation   *Synthesis* (1994), 445
betanes, heterocyclic mesomeris   *Tetrahedron* (1985), **41**, 2239
biaryls, Ullmann synthesis   *Synthesis* (1974), 9
bicyclic amidines   *Synthesis* (1972), 591

bicyclic lactams, chiral — *Tetrahedron* (1991), **47**, 9503
bicyclo[3.2.0]heptanones — *Synthesis* (1977), 155
bicyclo[3.3.1]nonanes — *Synthesis* (1979), 321
Biginelli dihydropyridine — *Tetrahedron* (1993), **49**, 6937
binding, dimensional probes — *Tetrahedron* (1986), **42**, 1917
bioconversion, of sesquiterpenes — *Tetrahedron* (1990), **46**, 4109
biohydroxylation, of terpenes — *Tetrahedron* (1984), **40**, 3597
biomimetic oxygenation — *Tetrahedron* (1977), **33**, 2869
biopolymers — *Tetrahedron* (1995), **51**, 8665
biosynthesis, alkaloid — *Tetrahedron* (1991), **47**, 5945
biosynthesis, allylic pyrophosphate — *Tetrahedron* (1980), **36**, 1109
biosynthesis, aromatic compounds — *Tetrahedron* (1978), **34**, 3353
biosynthesis, aromatic hemiterpenes — *Tetrahedron* (1978), **34**, 813
biosynthesis, β-lactam antibiotics — *Tetrahedron* (1977), **33**, 1545
biosysntheis of ergot alkaloids — *Tetrahedron* (1976), **32**, 873
biosynthesis, protein tRNA — *Tetrahedron* (1977), **33**, 1671
biosynthesis, sulfur compounds — *Tetrahedron* (1983), **39**, 1215
Birch reduction — *Synthesis* (1972), 391
bis-Wittig reactions — *Synthesis* (1975), 765
bismuthides, synthesis and reactions — *Synthesis* (1974), 328
bond formation with organoborons — *Synthesis* (1976), 633
boraheterocycles, synthesis — *Tetrahedron* (1977), **33**, 2331
borane, catechol — *Tetrahedron* (1976), **32**, 981
borohydrides, sulfurated — *Synthesis* (1972), 526
boron compounds, bond formation — *Synthesis* (1976), 633
boron compounds, chiral — *Tetrahedron* (1981), **37**, 3547
boron compounds, chiral — *Synthesis* (1986), 973
boron compounds, methanetetraboronic esters — *Synthesis* (1975), 147
boron compounds, rearrangements — *Synthesis* (1973), 635
boron compounds, thexylborane — *Synthesis* (1974), 77
boronic esters — *Synthesis* (1975), 147
buckminsterfullerene — *Synthesis* (1995), 895

C

cage molecules, polynitropolycyclic — *Tetrahedron* (1988), **44**, 2377
Calicheamicin — *Tetrahedron* (1994), **50**, 1397
calizarenes — *Tetrahedron* (1996), **52**, 2663
camphor derivatives — *Tetrahedron* (1987), **43**, 1969
camphor sultam, Oppolzer's — *Tetrahedron* (1993), **49**, 293
camptothecin — *Tetrahedron* (1981), **37**, 1047
carbanions, electroorganic — *Tetrahedron* (1993), **49**, 9627
carbanions, polyalkali derivatives — *Synthesis* (1977), 509

carbenes, heteroatom containing		*Tetrahedron* (1982), **38**, 2751
carbenes, insertion into O-H		*Tetrahedron* (1995), **51**, 10811
carbenium ions		*Tetrahedron* (1996), **52**, 6823
carbocation rearrangements		*Tetrahedron* (1976), **32**, 179
carbocations, multiply charged		*Tetrahedron* (1984), **40**, 4161
carbocycles,		*Tetrahedron* (1995), **51**, 9767
carbocyclic nucleosides, chiral		*Tetrahedron* (1992), **48**, 517
carbocyclic spiro compounds		*Synthesis* (1976), 425
carbodiimides		*Tetrahedron* (1981), **37**, 233
carbohydrate chemistry, enzyme		*Synthesis* (1991), 499
  catalysis
carbohydrate derivatives		*Tetrahedron* (1992), **48**, 2803
carbohydrates		*Synthesis* (1991), 499
carbohydrates, chiral intermediates		*Tetrahedron* (1984), **40**, 3161
carbohydrates, chiral templates		*Tetrahedron* (1991), **47**, 6079
carbohydrates, chiral synthon		*Tetrahedron* (1997), **53**, 14823
carbohydrates, cyclodextrins		*Tetrahedron* (1983), **39**, 1417
carbohydrates, from acyclic		*Tetrahedron* (1993), **49**, 5683
  precursors
carbohydrates, from siloxy dienes		*Synthesis* (1995), 607
carbohydrates, oxidation		*Synthesis* (1997), 597
carbon monoxide, carbonylation of		*Synthesis* (1985), 253
  organometallics
carbon disulfide, reactions		*Synthesis* (1984), 797
carbon–carbon bond formation		*Synthesis* (1992), 1185
carbonates, vinylene trihetero		*Tetrahedron* (1986), **42**, 1209
carbonic acid, ortho O,N-derivatives		*Synthesis* (1977), 73
carbonic ortho esters		*Synthesis* (1974), 153
carbonyl compounds, chiroptical		*Tetrahedron* (1986), **42**, 777
carbonyl compounds, α-fluoro		*Tetrahedron* (1985), **41**, 1111
carbonyl compounds, reductions		*Tetrahedron* (1986), **42**, 6351
carbonyl compounds, synthesis		*Synthesis* (1979), 633
carbonyl compounds, thio,		*Tetrahedron* (1985), **41**, 5393
  photochemistry
carbonyl compounds, unsaturated,		*Synthesis* (1975), 1
  cyclization
carbonyl groups, polarization		*Synthesis* (1977), 357
  reversal
carbonyl groups, transposition		*Tetrahedron* (1983), **39**, 345
carbonyl ylides		*Tetrahedron* (1976), **32**, 2165
carbonylation of organometallics		*Synthesis* (1985), 253
carbonylation reactions		*Synthesis* (1973), 509
carboxylic acids and esters		*Synthesis* (1973), 509

carboxylic ortho esters — *Synthesis* (1974), 153

carpyrinic acid — *Synthesis* (1972), 464

Cascade reactions — *Tetrahedron* (1996), **52**, 9289

catalysis, by arene chromium tricarbonyls — *Synthesis* (1993), 643

catalysis, by palladium — *Synthesis* (1992), 413

catalysis, by palladium — *Synthesis* (1992), 803

catalysis, by rhodium — *Tetrahedron* (1991), **47**, 1765

catalysis, by zinc — *Tetrahedron* (1992), **48**, 9577

catalysis, of H-migration — *Synthesis* (1991), 665

catalyst, enzyme — *Tetrahedron* (1978), **34**, 813

catalyst, epoxide hydrolase — *Tetrahedron* (1997), **53**, 15617

catalyst, palladium — *Synthesis* (1984), 369

catalyst, transition metal in carbonylation reactions — *Synthesis* (1973), 509

catalysts, hydroxy phosphines — *Synthesis* (1997), 983

catalysts, optically active transition metal — *Synthesis* (1988), 645

catalytic hydrogen transfer with ammonium formate — *Synthesis* (1988), 91

catalytic semihydrogenation of alkynes — *Synthesis* (1973), 457

catecholborane — *Tetrahedron* (1976), **32**, 981

cation-radical dimerization of olefins — *Synthesis* (1974), 539

cations, oxyallyl, synthesis with — *Tetrahedron* (1986), **42**, 4611

cavity-shaped molecules — *Tetrahedron* (1991), **47**, 6851

CEP method of preparation of phosphodiesters — *Synthesis* (1985), 449

ceric ion oxidation — *Synthesis* (1973), 347

chelators, iron — *Tetrahedron* (1995), **51**, 3939

chemical reactions, invention of — *Tetrahedron* (1992), **48**, 2529

chiral allenes — *Synthesis* (1973), 25

chiral auxiliaries, camphor — *Tetrahedron* (1987), **43**, 1969

chiral synthesis, amino alcohols — *Synthesis* (1982), 605

chiral synthesis with metal catalysts — *Synthesis* (1988), 645

chiral synthons, from carbohydrates — *Tetrahedron* (1984), **40**, 3161

chirality, and biological activity — *Tetrahedron* (1992), **48**, 8155

chirality, in anonists and antagonists — *Tetrahedron* (1991), **47**, 9953

chirality, isotopically engendered — *Tetrahedron* (1981), **37**, 4123

chirality multiplication — *Tetrahedron* (1994), **50**, 4259

3-chloro-2-aza-2-propeniminium — *Synthesis* (1988), 655

chlorocarbonyl isocyanate — *Tetrahedron* (1993), **49**, 3227 chloromethyltrimethylsilane — *Synthesis* (1985), 717

chlorosulfonyl isocyanate — *Synthesis* (1986), 437

chlorosulfides — *Tetrahedron* (1986), **42**, 3731

chorismic acid — *Synthesis* (1993), 179

chromium compounds — *Synthesis* (1992), 248

chromium II, salts — *Synthesis* (1974), 1

Cieplak model — *Tetrahedron* (1996), **52**, 5263

circular dichroism — *Tetrahedron* (1976), **32**, 2475

circular dichroism, cyclohexenones — *Tetrahedron* (1982), **38**, 3

circular dichroism, isotopically chiral compounds — *Tetrahedron* (1981), **37**, 4123

Claisen rearrangement — *Synthesis* (1977), 589

Claisen rearrangement, acetylenes in — *Tetrahedron* (1981), **37**, 3765

clay supported copper II, and iron III, nitrates — *Synthesis* (1985), 909

cleavage, reductive of nitro groups — *Synthesis* (1986), 693

cobalt complexes — *Tetrahedron* (1994), **50**, 1397

cobalt-mediated cyclization — *Tetrahedron* (1996), **52**, 4149

coelacanths — *Tetrahedron* (1994), **50**, 4235

cohalogenation — *Synthesis* (1993), 1177

combinatorial chemistry — *Tetrahedron* (1996), **52**, 4433

complexes, transition metal, in diyne reaction — *Synthesis* (1974), 761

conduritols — *Tetrahedron* (1990), **46**, 3715

conformational effect, problems — *Tetrahedron* (1977), **33**, 3193

conformational mobility, of organometallics — *Tetrahedron* (1993), **49**, 10175

conformations, hydroxylamine — *Tetrahedron* (1981), **37**, 849

conjugate addition by organocoppers — *Synthesis* (1985), 364

convergent solid-phase synthesis — *Tetrahedron* (1993), **49**, 11065

Cope rearrangement — *Tetrahedron* (1993), **49**, 5203

copper compounds — *Tetrahedron* (1984), **40**, 5005

copper II compounds, supported — *Synthesis* (1985), 909

copper compounds. copper-isonitrile complexes — *Synthesis* (1975), 291

copper compounds, enolate trapping — *Synthesis* (1985), 364

copper compounds, organocopper I — *Synthesis* (1972), 63

copper-isonitrile complexes, cyclic — *Synthesis* (1975), 291

copper II, nitrates, clay supported — *Synthesis* (1985), 909

Corey's reagent, dimethylsulfoxonium methylide — *Tetrahedron* (1987), **43**, 2609

corrins	*Tetrahedron* (1976), **32**, 1599
coumarins,	*Synthesis* (1982), 337
coupling reactions	*Tetrahedron* (1995), **51**, 8665
coupling reactions	*Synthesis* (1979), 633
coupling reactions, cross	*Synthesis* (1987), 1043
coupling reagents in peptides	*Synthesis* (1972), 453
cross-coupling reactions	*Synthesis* (1987), 1043
crown ethers	*Synthesis* (1976), 168
cyclopalladated complexes	*Synthesis* (1985), 233
crown ethers	*Synthesis* (1976), 168
crown compounds	*Tetrahedron* (1980), **36**, 461
cryptophanes	*Tetrahedron* (1987), **43**, 5725
cumulenes, hetero	*Tetrahedron* (1991), **47**, 1563
cuprates	*Synthesis* (1987), 325
cuprates, higher order	*Tetrahedron* (1984), **40**, 5005
cuprates	*Synthesis* (1972), 63
cyclic compounds from isonitrile-copper complexes	*Synthesis* (1975), 291
cyclic enediol phosphoryl synthesis of phosphodiesters	*Synthesis* (1985), 449
cyclic phosphorylating reagents	*Synthesis* (1993), 1
cyclic sulfides	*Tetrahedron* (1982), **38**, 2857
cyclic sulfites	*Synthesis* (1992), 1035
cyclic systems, silver ion catalyzed rearrangement	*Synthesis* (1975), 347
cyclic ureas by α-ureidoalkylation	*Synthesis* (1973), 243
cyclopentaannelation	*Synthesis* (1984), 529
cyclization of unsaturated carbonyls	*Synthesis* (1975), 1
cyclization of ylidenemalonodinitriles	*Synthesis* (1976), 705
Cyclization, palladium-catalyzed	*Tetrahedron* (1996), **52**, 9289
cyclization, radical	*Tetrahedron* (1997), **53**, 17543
cyclization reactions	*Tetrahedron* (1990), **46**, 1385
cyclizations, Nazarov	*Synthesis* (1983), 429
cycloaddition [4 + 4]	*Tetrahedron* (1996), **52**, 6251
cycloaddition, arene-dienophile	*Synthesis* (1980), 769
cycloaddition, Diels-Alder	*Tetrahedron* (1978), **34**, 19
cycloaddition, Diels-Alder	*Tetrahedron* (1992), **48**, 9111
cycloaddition, 1,3-dipolar	*Tetrahedron* (1991), **47**, 2925
cycloaddition, diradicals	*Tetrahedron* (1977), **33**, 3009
cycloaddition, of isomunchnones	*Synthesis* (1994), 123
cycloaddition, of o-quinodimethanes	*Synthesis* (1978), 793
cycloaddition reactions, acetylenes	*Tetrahedron* (1984), **40**, 2585

cycloaddition reactions, 1, 3-dipolar   *Synthesis* (1973), 71

cycloaddition reactions of acylisocyanates   *Synthesis* (1974), 461

cycloaddition reactions of tetracyanoethylene   *Synthesis* (1987), 749

cycloaddition reactions with azines and imines   *Synthesis* (1976), 349

cycloaddition reactions with sulfur   *Tetrahedron* (1988), **44**, 6755

cycloaddition reactions with cyclopropenes   *Synthesis* (1972), 675

cycloaddition, retro-   *Synthesis* (1977), 270

cycloaddition, spiro compounds   *Synthesis* (1978), 77

cycloadditions, cyclopropenes   *Synthesis* (1982), 701

cycloadditions, 1, 3-dipolar-, of nitrones   *Synthesis* (1975), 205

cycloadditions, intramolecular   *Tetrahedron* (1981), **37**, 3

cycloadditions, intramolecular   *Synthesis* (1995), 1205

cycloadditions, isocyanate   *Synthesis* (1982), 433

cycloadditions, steric control   *Synthesis* (1995), 1205

cycloadditions, to 1-azirines   *Synthesis* (1975), 483

cycloadditions with TCNE   *Synthesis* (1987), 749

cycloalkynes   *Synthesis* (1972), 599

cyclobutadiene   *Tetrahedron* (1980), **36**, 343

cyclobutanes,   *Synthesis* (1974), 539

cyclobutenedione   *Synthesis* (1978), 1

cyclobutenediones   *Synthesis* (1978), 649

cyclobutenediones   *Synthesis* (1978), 869

cyclodextrinbs   *Tetrahedron* (1983), **39**, 1417

cyclofunctionalization, of double bonds   *Tetrahedron* (1990), **46**, 3321

cyclohexanes, nucleophilic addition   *Tetrahedron* (1996), **52**, 5263

cyclohexylstatine   *Tetrahedron* (1996), **52**, 7063

cycloiminium salts, 2-substituted   *Synthesis* (1995), 361

cycloimmonium ylides   *Tetrahedron* (1976), **32**, 2647

cyclooctanoids   *Tetrahedron* (1996), **52**, 6251

cyclooctatetraene   *Tetrahedron* (1975), **31**, 2855

cyclooligomerization, of oxetane and oxirane   *Tetrahedron* (1993), **49**, 8707

cyclopalladated complexes   *Synthesis* (1985), 233

cyclopentenes, 3-oxo,   *Synthesis* (1973), 397

cyclophanes   *Tetrahedron* (1994), **50**, 4575

cyclopropanation, diastereoselective   *Tetrahedron* (1990), **46**, 4951

cyclopropanation/Cope
rearrangement, tandem *Tetrahedron* (1993), **49**, 5203
cyclopropane amino acids *Tetrahedron* (1990), **46**, 2231
cyclopropane, radical *Tetrahedron* (1981), **37**, 1625
cycloproparenes *Tetrahedron* (1988), **44**, 1305
cyclopropenes, cycloadditions *Synthesis* (1982), 701
cyclopropenes, cycloadditions *Synthesis* (1972), 675
cycloreversion, 4p+2p *Synthesis* (1985), 121
cyclotriveratrylenes *Tetrahedron* (1987), **43**, 5725
cysteine proteinases *Tetrahedron* (1976), **32**, 291

D
DBN *Synthesis* (1972), 591
DBU *Synthesis* (1972), 591
DEAD-TPP *Synthesis* (1981), 1
dealkoxycarbonylation *Synthesis* (1982), 805
dealkoxycarbonylation *Synthesis* (1982), 893
dealkylation, of amines *Synthesis* (1989), 1
dealkylation of ethers *Synthesis* (1988), 749
dehalogenation with sulfur to *Synthesis* (1985), 586
heterocycles
deltamethrinic acid, *cis*-1S, 3R,- *Tetrahedron* (1990), **46**, 4951
denitration reactions *Synthesis* (1988), 833
deprotection, of esters *Tetrahedron* (1993), **49**, 3691
deprotection, of silyl ethers *Synthesis* (1993), 11
deuterated alcohols *Synthesis* (1972), 254
diamonoid hydrocarbons *Tetrahedron* (1980), **36**, 971
dianion chemistry *Tetrahedron* (1991), **47**, 4223
dianions, of acids and ester enolates *Synthesis* (1982), 521
dianions, polyconjugated *Tetrahedron* (1988), **44**, 6957
diazabicyclo[4.3.0]non-5-ene *Synthesis* (1972), 591
diazabicyclo[5.4.0]unde-7-ene *Synthesis* (1972), 591
diazo compounds *Synthesis* (1972), 351
diazo-carbonyl compounds *Tetrahedron* (1991), **47**, 1765
diazoalkane substitution, *Synthesis* (1985), 569
electrophilic
diazoalkanes, cycloadditions *Tetrahedron* (1991), **47**, 2925
diazoketones *Tetrahedron* (1981), **37**, 2407
diazomalonic esters, reactions of *Synthesis* (1973), 137
Dibal, reduction with *Synthesis* (1975), 617
dicarbonyls, transition metal *Tetrahedron* (1996), **52**, 3377
complexes

dideoxy nucleosides   *Synthesis* (1992), 1
didehydroamino acids   *Synthesis* (1988), 159
didehydropeptides   *Synthesis* (1988), 159
dideoxynucleosides   *Synthesis* (1995), 1465
dienes, photooxidation   *Tetrahedron* (1991), **47**, 1343
Diels-Alder reaction   *Tetrahedron* (1978), **34**, 19
Diels-Alder reaction   *Synthesis* (1988), 569
Diels-Alder reaction   *Tetrahedron* (1992), **48**, 9111
Diels-Alder reaction   *Tetrahedron* (1996), **52**, 6251
Diels-Alder reaction, arenes   *Synthesis* (1980), 165
Diels-Alder reaction, azadienes   *Tetrahedron* (1983), **39**, 2869
Diels-Alder reaction, cycloreversion   *Synthesis* (1985), 121
Diels-Alder reaction, furans   *Tetrahedron* (1997), **53**, 14179
Diels-Alder reaction, heterodienes   *Synthesis* (1981), 753
Diels-Alder reaction,   *Tetrahedron* (1983), **39**, 3087
  heterodienophiles
Diels-Alder reaction, retro-   *Synthesis* (1977), 270
Diels-Alder reaction with TCNE   *Synthesis* (1987), 749
Diels-Alder reaction with   *Synthesis* (1986), 249
  tetracyanoethylene
Diels-Alder, retro   *Synthesis* (1987), 207
Diels-Alder with *o*-quinodimethanes   *Tetrahedron* (1987), **43**, 2873
dienes, hetero-substituted   *Synthesis* (1981), 753
dienes, silyl-substituted   *Synthesis* (1993), 349
dienophiles reactions with azines   *Synthesis* (1976), 349
  and imines
dienophiles, addition to arenes and   *Synthesis* (1980), 769
  heterocycles
difluoromethylenes, gem-, synthesis   *Tetrahedron* (1996), **52**, 8619
difunctionalization, in carbohydrates   *Tetrahedron* (1992), **48**, 2803
diisobutylaluminium hydride   *Synthesis* (1975), 617
dimerization of olefins   *Synthesis* (1974), 539
dimethylsulfoxide, activated   *Synthesis* (1981), 165
dimethylsulfoxide, oxidation with   *Synthesis* (1990), 857
dipolar cycloaddition   *Tetrahedron* (1991), **47**, 2925
dipolar cycloadditions, to   *Synthesis* (1973), 71
  heterocycles
dipolar cycloadditions of nitrones   *Synthesis* (1975), 205
dipoles   *Synthesis* (1991), 181
directionality of reactions   *Tetrahedron* (1983), **39**, 1013
dithianes   *Tetrahedron* (1989), **45**, 7643
diterpenoids   *Tetrahedron* (1995), **51**, 4571
dithiocaroxylic acids   *Synthesis* (1983), 605

dithiolonaphthothiazine	*Synthesis* (1985), 586
diyne reaction via transition metals	*Synthesis* (1974), 761
DMSO, activated	*Synthesis* (1981), 165
DNA strands, topological	*Tetrahedron* (1985), **41**, 3161
DNA synthesis	*Tetrahedron* (1983), **39**, 3
dodecahedrane	*Tetrahedron* (1979), **35**, 2189
donor-acceptor interactions	*Tetrahedron* (1987), **43**, 3839
drug design	*Tetrahedron* (1995), **51**, 8135
dyes	*Tetrahedron* (1991), **47**, 909
dyes, dithiolene	*Tetrahedron* (1991), **47**, 909
dyes, Stenhouse salt	*Tetrahedron* (1977), **33**, 463
Dynemicin	*Tetrahedron* (1994), **50**, 1397

E

electrochemistry, organometallic	*Synthesis* (1973), 377
electrocyclic reactions	*Synthesis* (1991), 181
electrolysis	*Tetrahedron* (1984), **40**, 811
electrolysis	*Tetrahedron* (1984), **40**, 935
electron transfer, activation induced	*Tetrahedron* (1990), **46**, 6193
electron transfer	*Synthesis* (1989), 233
electroorganic reactions	*Tetrahedron* (1984), **40**, 811
electroorganic, role of electroreductive mediators	*Synthesis* (1986), 873
electroorganic synthesis	*Tetrahedron* (1993), **49**, 9627
electrophilic diazoalkane substitution	*Synthesis* (1985), 569
electroreductive mediators	*Synthesis* (1986), 873
eliminations	*Tetrahedron* (1975), **31**, 2999
ellipticine alkaloids	*Tetrahedron* (1986), **42**, 2389
ellipticine, synthesis	*Synthesis* (1984), 289
enamides, photochemistry	*Synthesis* (1987), 421
enamides, photochemistry	*Synthesis* (1978), 489
enamines	*Tetrahedron* (1982), **38**, (1975)
enamines	*Tetrahedron* (1982), **38**, 3363
enamines, conjugated	*Tetrahedron* (1984), **40**, 2989
enantioface-differentiating reactions	*Tetrahedron* (1995), **51**, 8665
enantioselective synthesis	*Synthesis* (1988), 645
ene reactions, imine	*Synthesis* (1995), 347
enolate trapping reactions	*Synthesis* (1985), 364
enolates, ketone	*Tetrahedron* (1976), **32**, 2979
enolates, O-silyl	*Synthesis* (1977), 91
enolates, alkali metal	*Tetrahedron* (1977), **33**, 2737
enolization, photo	*Tetrahedron* (1976), **32**, 405

enzymatic catalyst   *Tetrahedron* (1978), **34**, 813
enzymatic compounds   *Synthesis* (1986), 605
enzymatic reactions   *Tetrahedron* (1986), **42**, 1917
enzyme mimetics   *Tetrahedron* (1995), **51**, 9241
enzyme mimetics   *Tetrahedron* (1995), **51**, 8665
enzyme catalysis, in carbohydrates   *Synthesis* (1991), 499
enzyme-catalyzed acyl transfer   *Synthesis* (1992), 895
enzymes   *Tetrahedron* (1986), **42**, 3351
enzymes   *Tetrahedron* (1991), **47**, 6003
enzymes, artificial   *Tetrahedron* (1984), **40**, 269
enzymes, artificial   *Tetrahedron* (1988), **44**, 5515
enzymes, esterolytic   *Synthesis* (1991), 1040
enzymes, for amino acid synthesis   *Tetrahedron* (1988), **44**, 5515
enzymes, in resolutions   *Tetrahedron* (1988), **44**, 1679
enzymes, in polysaccharide   *Tetrahedron* (1985), **41**, 2957
    modifications
enzymes, stereochemistry of   *Tetrahedron* (1982), **38**, 1541
    reactions
enzymes, suicide substrates   *Tetrahedron* (1982), **38**, 871
enzymology, of alkaloid   *Tetrahedron* (1991), **47**, 5945
    biosynthesis
eposide hydrolases   *Tetrahedron* (1997), **53**, 15617
epoxidation with peracids   *Tetrahedron* (1976), **32**, 2855
epoxidation, Sharpless   *Synthesis* (1986), 89
epoxides   *Tetrahedron* (1983), **39**, 2323
epoxides   *Synthesis* (1984), 629
epoxides, cheleselective reactions   *Synthesis* (1995), 745
ergoline ring, synthesis   *Tetrahedron* (1980), **36**, 3123
ergot alkaloids   *Tetrahedron* (1976), **32**, 873
Esperamicin   *Tetrahedron* (1994), **50**, 1397
esterification   *Tetrahedron* (1980), **36**, 2409
esterolytic enzymes   *Synthesis* (1991), 1040
esters, deprotection   *Tetrahedron* (1993), **49**, 3691
esters, diazomalonic, reactions of   *Synthesis* (1973), 137
esters from carbonylation reactions   *Synthesis* (1973), 509
esters, hydrolysis   *Tetrahedron* (1975), **31**, 2463
esters, ortho-, carboxylic and   *Synthesis* (1974), 153
    carbonic
ether cleavage   *Synthesis* (1983), 249
etherification reactions   *Tetrahedron* (1995), **51**, 10811
ethers, acetoxy   *Synthesis* (1988), 95
ethers, cleavage   *Tetrahedron* (1997), **53**, 13509
ethers, cleavage   *Synthesis* (1983), 249

ethers, crown — *Synthesis* (1976), 168
ethers, thio-, selective dealkylation — *Synthesis* (1988), 749
ethoxy carbonyl isothiocyanate — *Synthesis* (1975), 301
ethylene oxide, oligomers — *Tetrahedron* (1980), **36**, 461
ethylenic compounds, probes — *Tetrahedron* (1987), **43**, 3839
extrusion reactions — *Tetrahedron* (1988), **44**, 6241

F

flash photolysis — *Tetrahedron* (1996), **52**, 6823
flavanoids, oligomeric — *Tetrahedron* (1992), **48**, 1743
fluorides, alkali — *Synthesis* (1983), 169
fluorinated organometallics — *Tetrahedron* (1992), **48**, 189
fluorinated organometallics — *Tetrahedron* (1994), **50**, 2993
fluorinating agents — *Tetrahedron* (1991), **47**, 5329
fluorinating agents — *Tetrahedron* (1995), **51**, 6605
fluorination — *Tetrahedron* (1996), **52**, 8619
fluorine, introduction — *Tetrahedron* (1978), **34**, 3
fluorine compounds — *Tetrahedron* (1987), **43**, 3123
fluorine compounds — *Tetrahedron* (1991), **47**, 3207
fluorine compounds, polyfluoroaromatics, — *Synthesis* (1976), 652
fluorine compounds, from thermolytic reactions — *Synthesis* (1976), 374
fluorine compounds, carbonyls — *Tetrahedron* (1985), **41**, 1111
fluoroorganic compounds — *Synthesis* (1976), 374
fluoroorganics, chiral and bioactive — *Tetrahedron* (1993), **49**, 9385
fluvalenes, tetrahetero, — *Tetrahedron* (1986), **42**, 1209
formaldehyde, Prins reaction — *Synthesis* (1977), 661
formamide acetals — *Tetrahedron* (1979), **35**, 1675
forskolin, synthetic routes to — *Tetrahedron* (1992), **48**, 963
four-membered rings from isocyanides — *Synthesis* (1985), 1083
fragmentation radical reactions — *Synthesis* (1988), 417
fragmentation radical reactions — *Synthesis* (1988), 489
free radicals in telomerization — *Synthesis* (1977), 145
free radical functionalization — *Tetrahedron* (1995), **51**, 7095
Friedel-Crafts reactions, antimony pentahalides — *Synthesis* (1980), 345
Friedel-Crafts acylations — *Synthesis* (1972), 533
fullerenes — *Synthesis* (1995), 895
fullerenes — *Tetrahedron* (1993), **49**, 9207
fullerenes, thin film — *Tetrahedron* (1996), **52**, 5113
fulvalenes, tetrahetero — *Tetrahedron* (1986), **42**, 1209

fulvalenes, tetrathia-, preparation   *Synthesis* (1976), 489
functionalized polymers   *Synthesis* (1981), 413
furanose, glycosylation   *Synthesis* (1995), 1465

G
genetic engineering   *Tetrahedron* (1991), **47**, 6045
genetically engineered synthesis   *Tetrahedron* (1992), **48**, 2559
glutaraldehyde monoacetal   *Synthesis* (1985), 592
glyceraldehyde, R and S,   *Tetrahedron* (1986), **42**, 447
   2,3-*o*-isopropylidene
glycosidation, of sialic acid   *Tetrahedron* (1990), **46**, 5835
glycosides, C-   *Tetrahedron* (1992), **48**, 8545
carbohydrate derivatives   *Tetrahedron* (1993), **49**, 5683
glycosides, C-aryl   *Synthesis* (1994), 1
glycosides, C-   *Tetrahedron* (1992), **48**, 8545
glycosylation   *Synthesis* (1995), 1465
Grignard reagents, activation   *Tetrahedron* (1975), **31**, 2735
Grignard reagents, Cu-catalyzed   *Tetrahedron* (1984), **40**, 641

H
halides, hydrogenolysis   *Synthesis* (1980), 425
halohydrins   *Synthesis* (1994), 225
haloketones   *Tetrahedron* (1981), **37**, 2949
halosulfonium salts   *Tetrahedron* (1982), **38**, 2597
hard, soft, acid, base HSAB,   *Tetrahedron* (1985), **41**, 3
   organometallic reactions
Heck reaction   *Tetrahedron* (1997), **53**, 7371
helicating ligands   *Tetrahedron* (1992), **48**, 10013
hemiterpenes, aromatic biosynthesis   *Tetrahedron* (1978), **34**, 143
heteroatom-anions, in electroorganic   *Tetrahedron* (1993), **49**, 9627
   synthesis
hetero-Cope rearrangement   *Synthesis* (1989), 71
heteroatoms, bond formation   *Tetrahedron* (1996), **52**, 13265
heterocarbonates, tri, vinylene,   *Tetrahedron* (1986), **42**, 1209
heterocumulenes, isocyanates   *Synthesis* (1987), 525
heterocycles   *Tetrahedron* (1991), **47**, 7503
heterocycles   *Synthesis* (1993), 931
heterocycles   *Synthesis* (1995), 361
heterocycles   *Tetrahedron* (1996), **52**, 3057
heterocycles, aminoalkyl   *Synthesis* (1995), 879
heterocycles, 1-azirines,   *Synthesis* (1975), 483
   cycloadditions

heterocycles, bora   *Tetrahedron* (1977), **33**, 2331
heterocycles, carpyrinic acid   *Synthesis* (1972), 464
heterocycles, containing P-N-N   *Synthesis* (1978), 557
heterocycles, by dehalogenation with sulfur   *Synthesis* (1985), 586
heterocycles, double bond formation   *Synthesis* (1992), 413
heterocycles, from 1,3-dipolar cycloaddition reactions   *Synthesis* (1973), 71
heterocycles, from isoxazoles   *Synthesis* (1975), 20
heterocycles, from lactones and lactams   *Synthesis* (1972), 151
heterocycles, from phosphines   *Synthesis* (1974), 775
heterocycles, from primary nitros   *Synthesis* (1974), 613
heterocycles, H substitution in   *Synthesis* (1991), 103
heterocycles, hydroazulenes   *Synthesis* (1972), 517
heterocycles, hydroxyalkyl   *Synthesis* (1995), 879
heterocycles, indazoles   *Synthesis* (1978), 633
heterocycles, indolizine nucleus   *Synthesis* (1976), 209
heterocycles, isobenzofurans   *Tetrahedron* (1988), **44**, 2093
heterocycles, isocyanates in six-membered rings   *Synthesis* (1987), 525
heterocycles, isoquinoline alkaloids   *Synthesis* (1972), 657
heterocycles, β-lactames   *Tetrahedron* (1978), **34**, 1731
heterocycles, mercaptoalkyl   *Synthesis* (1995), 879
heterocycles, mesomeric, betaines   *Tetrahedron* (1985), **41**, 2239
heterocycles, nitrogen   *Tetrahedron* (1991), **47**, 9503
heterocycles, nitrogen   *Tetrahedron* (1991), **47**, 7503
heterocycles, nitrogen   *Tetrahedron* (1991), **47**, 9131
heterocycles, nitrogen-containing   *Tetrahedron* (1992), **48**, 9111
heterocycles, oligopyridines   *Synthesis* (1976), 1
heterocycles, oxazolones   *Synthesis* (1975), 749
heterocycles, phosphorus   *Synthesis* (1993), 931
heterocycles, pyridines   *Synthesis* (1972), 464
heterocycles, pyridines,   *Synthesis* (1976), 1
heterocycles, 2-pyrazolines   *Synthesis* (1985), 1028
heterocycles, pyrrolenines and pyrroles   *Synthesis* (1976), 281
heterocycles, pyrroles, thieno-   *Synthesis* (1985), 143
heterocycles, 3-substituted pyrroles   *Synthesis* (1985), 353
heterocycles, 3,5-pyrazolidinediones   *Synthesis* (1985), 1028
heterocycles, pyrrolizine   *Synthesis* (1987), 10
heterocycles, pyrrolopyrimidines   *Synthesis* (1974), 837
heterocycles, ring closures   *Tetrahedron* (1987), **43**, 5171

heterocycles, seven-membered   *Synthesis* (1988), 569
heterocycles, synthesis with   *Synthesis* (1975), 415
  benzofuroxan
heterocycles, synthesis, ring closure   *Tetrahedron* (1996), **52**, 13265
heterocycles, via cycloadditions to   *Synthesis* (1975), 483
  1-azirines
heterocyclic betaines   *Tetrahedron* (1977), **33**, 3203
heterocyclic compounds, from   *Tetrahedron* (1980), **36**, 2163
  o-aminoaldehydes
heterocyclic compounds, hetarynes   *Tetrahedron* (1982), **38**, 427
heterocyclic compounds, lithiation   *Synthesis* (1983), 957
heterocyclic compounds, meso-ionic   *Tetrahedron* (1982), **38**, 2965
heterocyclic intermediates   *Tetrahedron* (1990), **46**, 3321
heteroepines   *Synthesis* (1988), 569
heterofluvalenes, tetra   *Tetrahedron* (1986), **42**, 1209
heterogeneous catalysis, alumina   *Tetrahedron* (1997), **53**, 7999
heteropentalenes, mesomertic   *Tetrahedron* (1977), **33**, 3203
  betaines
heteroxyxles, isothiocyanates in   *Synthesis* (1987), 525
  six-membered rings
hexachlorodimethyl carbonate   *Synthesis* (1996), 553
hexamethylenetetramine   *Synthesis* (1979), 161
hexamethylenetetramine   *Synthesis* (1979), 61
high pressure   *Synthesis* (1985), 1
high pressure   *Synthesis* (1985), 999
high-pressure methods   *Tetrahedron* (1991), **47**, 8463
homoenolate anions   *Tetrahedron* (1983), **39**, 205
homogeneous catalysis   *Synthesis* (1981), 85
homologation reagents   *Synthesis* (1988), 95
homolytic addition, bromo addends   *Synthesis* (1977), 145
homolytic substitution   *Tetrahedron* (1996), **52**, 13265
host-guest complexes   *Tetrahedron* (1996), **52**, 2663
host-guest complexes   *Tetrahedron* (1995), **51**, 9241
host-guest complexes, carceplexes   *Tetrahedron* (1995), **51**, 3395
HSAB, organometallic reactions   *Tetrahedron* (1985), **41**, 3
hyda-hydroxylamino-oximes   *Synthesis* (1986), 704
hydrazine, reaction with pyrone   *Tetrahedron* (1977), **33**, 3183
hydrolysis of esters and amides   *Tetrahedron* (1975), **31**, 2463
hydride, tri-n-butyltin   *Synthesis* (1987), 665
hydride reduction   *Tetrahedron* (1979), **35**, 449
hydride reduction   *Tetrahedron* (1979), **35**, 567
hydrides, alkoxyalumino-   *Synthesis* (1972), 217
hydroazulene   *Synthesis* (1972), 517

hydroboration reagents *Tetrahedron* (1976), **32**, 981
hydroboration, cyclic *Tetrahedron* (1977), **33**, 2331
hydrocarbons, base catalyzed addition reactions *Synthesis* (1974), 309
hydrogen chloride, with acyl chlorides and nitriles *Synthesis* (1973), 189
hydrogen transfer reductions *Synthesis* (1988), 91
hydrogen bonding, replacement of *Synthesis* (1991), 103
hydrogenation *Synthesis* (1988), 91
hydrogenation, catalytic of alkynes *Synthesis* (1973), 457
hydrogenation, ionic *Synthesis* (1974), 633
hydrogenolysis of halides *Synthesis* (1980), 425
hydroperoxides, hydroxy *Synthesis* (1996), 179
hydroproline *Tetrahedron* (1996), **52**, 4527
hydrosilylation *Synthesis* (1997), 813
hydroxyl groups, via organotin *Tetrahedron* (1985), **41**, 643
hydroxylamine, mesitylene O-sulfonic acid *Synthesis* (1977), 1
hydroxylamine, reaction with pyrones *Tetrahedron* (1977), **33**, 3183
hydroxylamines, conformations *Tetrahedron* (1981), **37**, 849
hydroxylation, acyl, at carbon, with metal salts *Synthesis* (1973), 567
hydroxylation, photochemical, of aromatics *Synthesis* (1974), 173
hydroxylation, substitutive acyloxylation at carbon *Synthesis* (1972), 1
hydrozones, aryl *Tetrahedron* (1980), **36**, 161
hypobromites *Tetrahedron* (1976), **32**, 517

I

imides *Synthesis* (1988), 655
imides, alkenylation of *Synthesis* (1975), 685
imides, dichloroalkyl *Synthesis* (1972), 599
imides, phosphine, for heterocycles *Synthesis* (1974), 775
imidic quaternary salts *Synthesis* (1988), 655
imines, azomethine *Synthesis* (1973), 469
imines, ene reactions *Synthesis* (1995), 347
imines, reaction with dienophiles *Synthesis* (1976), 349
iminium salts, 3-chloro-2-propene *Synthesis* (1979), 241
iminium salts, photochemistry *Tetrahedron* (1983), **39**, 3845
iminophosphoranes *Synthesis* (1973), 469
iminosulfuranes, Ei reactions *Tetrahedron* (1977), **33**, 2359

immonium ylides   *Tetrahedron* (1976), **32**, 2647

immunophilins   *Tetrahedron* (1992), **48**, 2545

inclusion complexes   *Tetrahedron* (1995), **51**, 8665

indazoles   *Synthesis* (1978), 633

indolizine nucleus   *Synthesis* (1976), 209

industrial asymmetric synthesis   *Tetrahedron* (1994), **50**, 3639

infrared absorption, near-, by   *Tetrahedron* (1991), **47**, 909
   dithiolene dyes

insect pheromones   *Tetrahedron* (1977), **33**, 1845

insect pheromones   *Synthesis* (1978), 413

insertion reactions, nitrogen   *Tetrahedron* (1981), **37**, 1283

intramolecular addition reactions of   *Synthesis* (1988), 263
   silanes

intramolecular alkylation to spiros   *Synthesis* (1974), 383

intramolecular interactions   *Tetrahedron* (1995), **51**, 8665

intramolecular reactions,   *Tetrahedron* (1985), **41**, 4367
   N-acyliminium intermediates

intramolecular $S_{N'}$ reaction   *Tetrahedron* (1992), **48**, 7383

invention of chemical reactions   *Tetrahedron* (1992), **48**, 2529

iodination   *Synthesis* (1988), 923

iodine compounds   *Synthesis* (1984), 709

iodine, hypervalent   *Synthesis* (1990), 431

iodoaromatic compounds   *Synthesis* (1988), 923

ion-radical reactions   *Tetrahedron* (1985), **41**, 2771

ionic hydrogenation   *Synthesis* (1974), 633

ions, nonclassical   *Tetrahedron* (1976), **32**, 179

Iridoids, lactone   *Tetrahedron* (1997), **53**, 14507

iron compounds, clay supported   *Synthesis* (1985), 909
   iron III, nitrates

iron compounds   *Synthesis* (1989), 341

iron compounds, aromatic   *Tetrahedron* (1983), **39**, 4027

iron III, nitrates, clay supported   *Synthesis* (1985), 909

isobenzofurans   *Tetrahedron* (1988), **44**, 2093

isocyanate, chlorosulfonyl   *Synthesis* (1986), 437

isocyanates   *Synthesis* (1987), 525

isocyanates, acyl additions of   *Synthesis* (1982), 433

isocyanates, acyl, cycloadditions   *Synthesis* (1974), 461

isocyanates, α-haloalkyl   *Synthesis* (1980), 85

isocyanato-phosphine derivatives   *Synthesis* (1986), 793

isocyanides, four-membered rings   *Synthesis* (1985), 1083

isomerization of azido-tetrazolos   *Synthesis* (1973), 123

isomerization, photochemical and   *Tetrahedron* (1976), **32**, 641
   thermal

isomunchnones, cycloaddition   *Synthesis* (1994), 123
isonitrile-copper complexes   *Synthesis* (1975), 291
isoprenoids, biosynthesis   *Tetrahedron* (1978), **34**, 143
isopropylideneglyceraldehyde,   *Tetrahedron* (1986), **42**, 447
    R and S,
isoquinoline alkaloids, by phenolic   *Synthesis* (1972), 657
    oxidation
isoquinolines,   *Tetrahedron* (1996), **52**, 4433
    4-aryl-1,2,3,4-tetrahydro
isothiocyanantes   *Synthesis* (1987), 525
isothiocyanate, ethoxy carbonyl   *Synthesis* (1975), 301
isoureas, reactions   *Synthesis* (1979), 561
isoxazoles   *Synthesis* (1987), 857
isoxazoles, heterocycles from   *Synthesis* (1975), 20

K

ketenes, addition reactions   *Tetrahedron* (1986), **42**, 2587
ketenes, halogenated   *Tetrahedron* (1981), **37**, 2949
ketenes, α-oxo, dithioacetals   *Tetrahedron* (1986), **42**, 3029
ketenes, thio   *Tetrahedron* (1988), **44**, 1827
ketone enolates   *Tetrahedron* (1976), **32**, 2979
ketone equivalents   *Synthesis* (1977), 289
ketones   *Tetrahedron* (1995), **51**, 9767
ketones, alkylation of   *Synthesis* (1983), 517
ketones, bridged bicyclic, ring   *Tetrahedron* (1987), **43**, 3
    expansions of
ketones, bridged bicyclic   *Tetrahedron* (1981), **37**, 2697
ketones, cheleselective reactions   *Synthesis* (1995), 745
ketones, fluorinated   *Tetrahedron* (1991), **47**, 3207
ketones, ORD and CD   *Tetrahedron* (1986), **42**, 777
ketones, reduction with alkali metals   *Tetrahedron* (1986), **42**, 6351
ketones, steroidal, reduction   *Synthesis* (1972), 526
ketones, unsaturated, reduction   *Tetrahedron* (1986), **42**, 6351
kinetics, of radical reactions   *Tetrahedron* (1993), **49**, 1151

L

lactam acetals   *Tetrahedron* (1988), **44**, 5975
β-lactams   *Synthesis* (1973), 327
β-lactams   *Tetrahedron* (1978), **34**, 1731
lactams   *Tetrahedron* (1995), **51**, 9767
lactams, chiral bicyclic,   *Tetrahedron* (1990), **46**, 4951
    cyclopropanation of

lactams, bicyclic, chiral — *Tetrahedron* (1991), **47**, 9503
β-lactams, reactions — *Synthesis* (1975), 547
lactams, synthesis by N-insertion — *Tetrahedron* (1981), **37**, 1283
lactams, synthesis of heterocycles — *Synthesis* (1972), 151
β-lactam, with organometallics — *Tetrahedron* (1988), **44**, 5615
lactones — *Synthesis* (1995), 729
lactones — *Tetrahedron* (1995), **51**, 9767
β-lactones — *Synthesis* (1993), 441
lactones, iridoid — *Tetrahedron* (1997), **53**, 14507
lactones, macrocyclic synthesis — *Tetrahedron* (1977), **33**, 3041
lactones, α-methylene, — *Synthesis* (1986), 157
lactones, α-methylene, — *Synthesis* (1975), 67
lactones, synthesis of heterocycles — *Synthesis* (1972), 151
Langmuir-Blodgett films — *Tetrahedron* (1996), **52**, 5113
lanthanides — *Tetrahedron* (1986), **42**, 6573
Lawesson's reagents, thioation — *Tetrahedron* (1985), **41**, 5061
lead compounds, lead IV, acetate and azides — *Synthesis* (1972), 285
leukotrienes — *Tetrahedron* (1983), **39**, 1687
Lewis acids, in reactions of allylic tin compounds — *Tetrahedron* (1993), **49**, 7395
ligands, helicating — *Tetrahedron* (1992), **48**, 10013
lignans, asymmetric synthesis — *Tetrahedron* (1990), **46**, 5029
lipase, transesterification catalysis — *Tetrahedron* (1996), **52**, 3769
lipolytic enzymes — *Synthesis* (1991), 1040
lipstatin — *Synthesis* (1995), 729
liquid crystals — *Tetrahedron* (1988), **44**, 3413
lithiation of heterocycles — *Synthesis* (1983), 957
lithium aluminium hydride — *Synthesis* (1972), 217
lithium aluminium hydride — *Synthesis* (1974), 691
lithium reagents — *Synthesis* (1975), 83
lithium/amine reduction — *Synthesis* (1972), 391

M
macrocycles — *Tetrahedron* (1995), **51**, 9767
macrocyclic compounds — *Tetrahedron* (1988), **44**, 1573
macrocyclic polyamines — *Tetrahedron* (1992), **48**, 6175
macrolide antibiotics — *Tetrahedron* (1985), **41**, 3569
macrolides, synthesis — *Tetrahedron* (1977), **33**, 3041
macrolides, synthesis — *Tetrahedron* (1977), **33**, 683
magnesium compounds, Barbier reaction — *Synthesis* (1977), 18
magnetic circular dichroism — *Tetrahedron* (1984), **40**, 3845

malonaldehyde monoacetal — *Synthesis* (1985), 592

malonic esters, diazo-, reactions of — *Synthesis* (1973), 137

malonitrile, derivatives — *Synthesis* (1981), 925

malonodinitriles, ylidene- — *Synthesis* (1976), 705

malononitrile — *Synthesis* (1978), 241

malononitrile — *Synthesis* (1978), 165

manganese dioxide in oxidations — *Synthesis* (1976), 65

manganese dioxide in oxidations — *Synthesis* (1976), 133

manganese III — *Synthesis* (1993), 833

Mannich bases — *Synthesis* (1973), 703

Mannich bases — *Tetrahedron* (1990), **46**, 1791

Mannich reaction — *Synthesis* (1984), 85

Mannich reaction — *Synthesis* (1984), 181

marine natural products — *Tetrahedron* (1977), **33**, 1421

mass spectrometry, filed desorption — *Tetrahedron* (1982), **38**, 1125

mass spectrometry, stereochemistry — *Tetrahedron* (1980), **36**, 2687

mechanism, diradical, dipolar cycloaddition — *Tetrahedron* (1977), **33**, 3009

mechanism, $S_{RN}1$-$S_{RN}2$ — *Tetrahedron* (1993), **49**, 4485

mechanisms, reactivity-selectivity principle — *Tetrahedron* (1980), 367, 3461

medium rings — *Tetrahedron* (1993), **49**, 10749

meralonic acid — *Tetrahedron* (1986), **42**, 3

mesoionic, heteropentalenes — *Tetrahedron* (1977), **33**, 3203

mesoionic compounds — *Synthesis* (1994), 123

metabolites, oxidized, of polycyclic aromatics — *Synthesis* (1986), 605

metal salts, in acyloxylation — *Synthesis* (1973), 567

metal reductions, Birch conditions — *Synthesis* (1972), 391

metal catalyzed dimerization — *Synthesis* (1974), 539

metallation, poly — *Tetrahedron* (1983), **39**, 2733

metallocene catalysts, chiral — *Tetrahedron* (1996), **52**, 12853

metals, alkali, reductions with — *Tetrahedron* (1986), **42**, 6351

metals, transition — *Synthesis* (1976), 561

α-methylene lactones, — *Synthesis* (1975), 67

α-methylenelactones, — *Synthesis* (1986), 157

methane tetraboronic esters — *Synthesis* (1975), 147

methyleneamino-phosphine — *Synthesis* (1986), 793

methyleniminium salts — *Tetrahedron* (1992), **48**, 3659

mevinic acids — *Tetrahedron* (1986), **42**, 4909

micellar systems, photochemistry — *Tetrahedron* (1982), **38**, 2455

microwaves — *Tetrahedron* (1995), **51**, 10403

Mitsunobu reaction — *Synthesis* (1981), 1

Modius strip approach   *Tetrahedron* (1985), **41**, 3161
molecular recognition   *Tetrahedron* (1995), **51**, 9241
monoterpenes, iridoid   *Tetrahedron* (1997), **53**, 14507

N
Nafion-H   *Synthesis* (1986), 513
naphthobisthiazine   *Synthesis* (1985), 586
natural products, annonaceus   *Synthesis* (1995), 1447
   acetogenins
natural products, from octocorals   *Tetrahedron* (1995), **51**, 4571
natural products, marine   *Tetrahedron* (1977), **33**, 1421
natural products, polyketide   *Tetrahedron* (1977), **33**, 2159
natural products synthesis,   *Tetrahedron* (1996), **52**, 6251
   cycloaddition
natural products, synthesis   *Tetrahedron* (1992), **48**, 2559
near-IR absorption, by dithiolenes   *Tetrahedron* (1991), **47**, 909
neocarzinostatin   *Tetrahedron* (1994), **50**, 1397
neucleosides, S- and Se-modified   *Tetrahedron* (1993), **49**, 9877
neurotoxins   *Tetrahedron* (1996), **52**, 6025
ninhydrin   *Tetrahedron* (1978), **34**, 1285
ninhydrin and analogs   *Tetrahedron* (1991), **47**, 8791
nitrenium ions   *Tetrahedron* (1996), **52**, 6823
nitrile, iso-, copper complexes   *Synthesis* (1975), 291
nitriles, in prebiotic synthesis   *Tetrahedron* (1984), **40**, 37
nitriles, in prebiotic synthesis   *Tetrahedron* (1984), **40**, 13
nitriles, 3-oxo   *Synthesis* (1984), 1
nitriles, reaction with acyl chlorides   *Synthesis* (1973), 189
nitrilium salts   *Tetrahedron* (1980), **36**, 1279
nitro compounds, for heterocycles   *Synthesis* (1974), 613
nitro groups, reductive cleavage   *Synthesis* (1986), 693
nitro groups, displacement of   *Synthesis* (1991), 423
nitro groups, substitution of   *Tetrahedron* (1978), **34**, 2057
nitroacetic acid   *Synthesis* (1979), 666
nitroalkanes as alkyl anion synthons   *Synthesis* (1988), 833
nitroalkanes, reduction   *Synthesis* (1995), 1053
nitroenamines   *Tetrahedron* (1981), **37**, 1453
nitrogen heterocycles, medium ring   *Tetrahedron* (1991), **47**, 9131
nitrogen heterocycles   *Tetrahedron* (1992), **48**, 9111
nitrogen bond cleavage   *Tetrahedron* (1980), **36**, 161
nitrogen compounds, azines   *Tetrahedron* (1988), **44**, 1
nitroketones   *Synthesis* (1980), 261
nitrones, 1, 3-dipolar cycloadditions   *Synthesis* (1975), 205

NMR of polyconjugated dianions — *Tetrahedron* (1988), **44**, 6957
NMR structural correlations, H/C — *Tetrahedron* (1991), **47**, 3521
nonbenzenoid aromatic ring systems — *Synthesis* (1975), 765
nonclassical ions — *Tetrahedron* (1976), **32**, 179
Nozaki-Hiyama reaction — *Synthesis* (1992), 248
nucleophilic acylation — *Tetrahedron* (1976), **32**, 1943
nucleophilic addition, to unsaturated ligands — *Tetrahedron* (1978), **34**, 3047
nucleophilic displacement, of nitros — *Synthesis* (1991), 423
nucleophilic substitution, aromatic — *Tetrahedron* (1978), **34**, 2057
nucleosides — *Synthesis* (1995), 1465
nucleosides, carbocyclic, chiral — *Tetrahedron* (1992), **48**, 517
nucleosides, dideoxy, synthesis — *Synthesis* (1992), 1
nucleotide, oligo-, synthesis — *Synthesis* (1975), 222
nucleotides, synthesis — *Tetrahedron* (1978), **34**, 3143

O

olefins, chiral — *Tetrahedron* (1976), **32**, 2475
olefins, inversion — *Tetrahedron* (1980), **36**, 557
olefins, Pd-catalyzed oxidation — *Synthesis* (1984), 369
olefins, radical addition — *Tetrahedron* (1980), **36**, 701
oligomeric flavanoids — *Tetrahedron* (1992), **48**, 1743
oligonucleotide synthesis — *Synthesis* (1975), 222
oligonucleotides — *Tetrahedron* (1993), **49**, 6123
oligonucleotides — *Synthesis* (1975), 222
oligonucleotides, functionalization — *Tetrahedron* (1993), **49**, 1925
oligopyridines — *Synthesis* (1976), 1
oligopyridines — *Tetrahedron* (1992), **48**, 10013
oligosaccharides — *Tetrahedron* (1996), **52**, 1095
Oppolzer's camphor sultam — *Tetrahedron* (1993), **49**, 293
optical resolution, strategies — *Tetrahedron* (1977), **33**, 2725
optical storage, reversible — *Tetrahedron* (1993), **49**, 8267
optically active compounds — *Tetrahedron* (1991), **47**, 4789
optically active transition metal catalysts — *Synthesis* (1988), 645
orbital interactions, secondary — *Tetrahedron* (1983), **39**, 2095
organoaluminiums — *Tetrahedron* (1988), **44**, 5001
organoborane reagents, chiral — *Tetrahedron* (1981), **37**, 3547
organoboranes chiral — *Synthesis* (1986), 973
organochlorozirconocene complexes — *Tetrahedron* (1996), **52**, 12853
organocopper chemistry, transmetallation — *Synthesis* (1993), 537

organocopper conjugate additions, with enolate trapping	*Synthesis* (1985), 364
organocuprates	*Synthesis* (1972), 63
organocuprates, higher-order mixed	*Synthesis* (1987), 325
organofluorine compounds	*Tetrahedron* (1987), **43**, 3123
organolithium reagents	*Synthesis* (1975), 83
organomercurials	*Tetrahedron* (1982), **38**, 1713
organometallic compounds	*Tetrahedron* (1984), **40**, 641
organometallic compounds, α-hetero-substituted	*Tetrahedron* (1980), **36**, 2409
organometallic compounds, cyclopalladated complexes	*Synthesis* (1985), 233
organometallic compounds, iron compounds	*Synthesis* (1989), 341
organometallic compounds, MO theory	*Tetrahedron* (1982), **38**, 1339
organometallic compounds, non-transition	*Tetrahedron* (1978), **34**, 3827
organometallic electrochemistry	*Synthesis* (1973), 377
organometallic reactions, chemoselectivity	*Tetrahedron* (1985), **41**, 3
organometallic reagents	*Synthesis* (1992), 803
organometallic reagents	*Tetrahedron* (1992), **48**, 189
organometallic reagents, carbonylation	*Synthesis* (1985), 253
organometallic reagents, tellurium	*Synthesis* (1986), 1
organometallics	*Tetrahedron* (1996), **52**, 7201
organometallics, and chirality	*Tetrahedron* (1994), **50**, 4259
organometallics, cuprates	*Synthesis* (1987), 325
organometallics, fluorinated	*Tetrahedron* (1992), **48**, 189
organometallics in β-lactam	*Tetrahedron* (1988), **44**, 5615
organometallics, reaction with tetracyanoethylene	*Synthesis* (1987), 959
organopalladium compounds	*Tetrahedron* (1977), **33**, 2615
organosilicon compounds	*Synthesis* (1979), 761
organosilicon compounds, oxidation	*Tetrahedron* (1996), **52**, 7599
organosilicon protecting groups	*Synthesis* (1985), 817
organothiophosphorus reagents	*Tetrahedron* (1985), **41**, 2567
organotin compounds	*Synthesis* (1992), 803
organotin compounds	*Synthesis* (1996), 423
organotin compounds, hydroxyl manipulation	*Tetrahedron* (1985), **41**, 643
organotransition metal reagents	*Synthesis* (1995), 745

organozinc reagents — *Tetrahedron* (1992), **48**, 9577

organozinc reagents, activation — *Tetrahedron* (1987), **43**, 2203

organozirconium compounds — *Synthesis* (1988), 1

organozirconium complexes — *Tetrahedron* (1995), **51**, 4255

orthoesters, carboxylic and carbonic — *Synthesis* (1974), 153

orthosomycins — *Tetrahedron* (1979), 35, 1207

oxazines, 1,3- — *Synthesis* (1972), 333

oxazoles, 1,2- — *Synthesis* (1975), 20

oxazolines, 2- — *Tetrahedron* (1994), **50**, 2297

oxazolines — *Tetrahedron* (1985), **41**, 837

oxaziridines, aminations with — *Synthesis* (1991), 327

oxazolones — *Synthesis* (1975), 749

oxetanes — *Synthesis* (1995), 729

oxidation — *Synthesis* (1996), 179

oxidation — *Synthesis* (1995), 1325

oxidation, C–Si bond — *Tetrahedron* (1996), **52**, 7599

oxidation catalysts, Zr alkoxide — *Synthesis* (1997), 1115

oxidation, of thioethers — *Tetrahedron* (1986), **42**, 5459

oxidation, of thioethers — *Tetrahedron* (1988), **44**, 6537

oxidation, of α,β-unsaturated ketones — *Synthesis* (1992), 235

oxidation of phenol — *Synthesis* (1972), 657

oxidation, olefins to ketones — *Synthesis* (1984), 369

oxidation, permanganate — *Synthesis* (1987), 85

oxidation, peroxidases — *Tetrahedron* (1997), **53**, 13183

oxidation with ceric ion — *Synthesis* (1973), 347

oxidation with chromium II, salts — *Synthesis* (1974), 1

oxidation, with hypervalent iodine — *Synthesis* (1990), 431

oxidation with manganese dioxide — *Synthesis* (1976), 65

oxidation with manganese dioxide — *Synthesis* (1976), 133

oxidation, with PCC — *Synthesis* (1982), 245

oxidation with periodic acid — *Synthesis* (1974), 229

oxidation with permanganate ion — *Synthesis* (1987), 85

oxidative deprotection, of silyl ethers — *Synthesis* (1993), 11

oxidized metabolites of polycyclic aromatics — *Synthesis* (1986), 605

oxidizing agents, sodium perborate — *Tetrahedron* (1995), **51**, 6145

oximmes, α-hydroxylamino — *Synthesis* (1986), 704

oxiranes, cleavage with metal halides — *Synthesis* (1994), 225

oxirans — *Tetrahedron* (1983), **39**, 2323

oxocarbons   *Synthesis* (1980), 961
3-oxocyclopentenes   *Synthesis* (1973), 397
α-oxoketene acetals   *Tetrahedron* (1990), **46**, 5423
α-oxoketene dithioacetals   *Tetrahedron* (1986), **42**, 3029
1-oxo-steroids, reduction   *Tetrahedron* (1995), **51**, 5711
oxy-Cope rearrangement   *Tetrahedron* (1997), **53**, 13971
oxyallyl cations   *Tetrahedron* (1986), **42**, 4611
oxygen insertion, in ketones   *Tetrahedron* (1981), **37**, 2697
oxygenation, biominetic   *Tetrahedron* (1977), **33**, 2869
oxyphosphoranes   *Synthesis* (1974), 90

P
palladium compounds   *Tetrahedron* (1977), **33**, 2615
palladium compounds   *Tetrahedron* (1986), **42**, 4361
palladium compounds, cyclopalladated complexes   *Synthesis* (1985), 233

palladium, in olefin reactions   *Tetrahedron* (1984), **40**, 2415
palladium salts in aromatic substitution of olefins   *Synthesis* (1973), 524

penicillanic acid   *Tetrahedron* (1975), **31**, 2321
penicillin, biosynthesis   *Tetrahedron* (1977), **33**, 1545
peptide, cyclic   *Tetrahedron* (1975), **31**, 2177
peptide mimetics   *Tetrahedron* (1997), **53**, 12798
peptide synthesis, azide method   *Synthesis* (1974), 549
peptide synthesis, solid-phase fragment condensation in   *Synthesis* (1994), 337

peptide synthesis   *Synthesis* (1972), 453
peptides by the azide method   *Synthesis* (1974), 549
peptides, didehydro   *Synthesis* (1988), 159
peptides, side reactions   *Synthesis* (1981), 333
peptides, solid-phase synthesis   *Tetrahedron* (1993), **49**, 11065
peptides with coupling reagents   *Synthesis* (1972), 453
peptidomemetic design   *Tetrahedron* (1993), **49**, 3547
peracids, epoxidation   *Tetrahedron* (1976), **32**, 2855
perfluorinated resinsulfonic acid Nafion-H   *Synthesis* (1986), 513

perfluoroalkyl organometallics   *Tetrahedron* (1992), **48**, 189
perfluoroslkanesulfonic esters   *Synthesis* (1982), 85
perhalogenated sulfenes thione oxides and dioxides,   *Synthesis* (1988), 349

perhydrates   *Synthesis* (1996), 179
pericyclic reactions   *Synthesis* (1988), 569

pericyclic reactions, diradical mechanism — *Tetrahedron* (1977), **33**, 3009
periodic acid, applications — *Synthesis* (1974), 229
permanganate oxidation — *Synthesis* (1987), 85
peroxides, acyloxylation — *Synthesis* (1972), 1
phanes — *Synthesis* (1973), 85
phenolic oxidation of alkaloids — *Synthesis* (1972), 657
phenols, oxidation — *Synthesis* (1997), 1115
pheromones, chiral — *Synthesis* (1978), 413
pheromones, sex, insect — *Tetrahedron* (1977), **33**, 1845
pheromones, synthesis — *Synthesis* (1977), 817
phosphates, enzymatic reactions — *Tetrahedron* (1982), **38**, 1541
phosphazines — *Tetrahedron* (1981), **37**, 437
phosphine compounds, for heterocycles — *Synthesis* (1974), 775
phosphine reactions with unsaturated compounds — *Synthesis* (1986), 793
phosphines, hydroxy — *Synthesis* (1997), 983
phosphodiesters, — *Synthesis* (1985), 449
phosphonates, halomethyl — *Synthesis* (1997), 727
phosphonates — *Tetrahedron* (1997), **53**, 16609
phosphonates, vinyl — *Synthesis* (1992), 333
phosphonium salts, vinyl-, for heterocycles — *Synthesis* (1974), 775
phosphoramidate approach — *Tetrahedron* (1993), **49**, 6123
phosphoramidite method — *Tetrahedron* (1993), **49**, 10441
phosphoranes, alkylidene-, for heterocycles — *Synthesis* (1974), 775
phosphorus compounds — *Synthesis* (1978), 557
phosphorus compounds, additions — *Synthesis* (1979), 81
phosphorus compounds, intramolecular reactions — *Tetrahedron* (1980), **36**, 1717
phosphorus compounds, organothio — *Tetrahedron* (1985), **41**, 2567
phosphorus compounds, oxyphosphoranes — *Synthesis* (1974), 90
phosphorus compounds, phosphodiesters, — *Synthesis* (1985), 449
phosphorus compounds, spin labeled — *Synthesis* (1981), 682
phosphorus compounds, stereochemistry — *Tetrahedron* (1980), **36**, 2059
phosphorus compounds, with unsaturated compounds — *Synthesis* (1986), 793
phosphorus heterocycles — *Synthesis* (1993), 931

phosphorus ylides   *Tetrahedron* (1996), **52**, 1855
phosphorylated molecules   *Tetrahedron* (1993), **49**, 10441
phosphorylating reagents, cyclic   *Synthesis* (1993), 1
phosphorylation methods,   *Synthesis* (1977), 737
phosphotriesters   *Tetrahedron* (1978), **34**, 3143
photochemical dimerization   *Synthesis* (1974), 539
photochemical hydroxylation of   *Synthesis* (1974), 173
  aromatics
photochemical isomerizations   *Tetrahedron* (1976), **32**, 641
photochemical reactivity   *Tetrahedron* (1992), **48**, 3251
photochemical synthesis   *Synthesis* (1989), 145
photochemical synthesis   *Synthesis* (1989), 233
photochemistry, benzene   *Tetrahedron* (1976), **32**, 1309
photochemistry, benzene   *Tetrahedron* (1977), **33**, 2459
photochemistry, enamides   *Synthesis* (1978), 489
photochemistry, enamides   *Synthesis* (1987), 421
photochemistry, iminium salts   *Tetrahedron* (1983), **39**, 3845
photochemistry, in organized media   *Tetrahedron* (1986), **42**, 5753
photochemistry, in micelles   *Tetrahedron* (1982), **38**, 2455
photochemistry, metal catalyzed   *Tetrahedron* (1983), **39**, 485
photochemistry, photosensitization   *Synthesis* (1981), 249
photochemistry, protective group   *Synthesis* (1980), 1
  removal
photochemistry, supramolecular   *Tetrahedron* (1992), **48**, 10443
photochemistry, thiocarbonyls   *Tetrahedron* (1985), **41**, 5393
photochemistry, vitamin A   *Tetrahedron* (1984), **40**, 1931
photocycloadditions   *Synthesis* (1980), 165
photoenolization   *Tetrahedron* (1976), **32**, 405
photoinduced electron transfer,   *Tetrahedron* (1994), **50**, 575
  radicals by
photooxidation, of 1,3-dienes   *Tetrahedron* (1991), **47**, 1343
photosensitive artificial membranes   *Tetrahedron* (1994), **50**, 4039
photosensitization   *Synthesis* (1981), 249
phthalide anions   *Tetrahedron* (1995), **51**, 5207
phytosiderophores   *Tetrahedron* (1995), **51**, 3939
pig liver esterases PLE   *Tetrahedron* (1996), **52**, 3769
pig liver esterase   *Tetrahedron* (1990), **46**, 6587
pigments, amanita muscaria   *Tetrahedron* (1979), 35, 2843
polyols, 1,3-, stereoselective   *Synthesis* (1991), 635
polarization reversal in carbonyls   *Synthesis* (1977), 357
polyalkali metal derivatives,   *Synthesis* (1977), 509
  heterofunctional compounds
polyamines, macrocyclic   *Tetrahedron* (1992), **48**, 6175

polyaza compounds	*Tetrahedron* (1991), **47**, 6851
polycarbenes	*Tetrahedron* (1995), **51**, 11337
polycarbocyclic sesquiterpenes	*Tetrahedron* (1985), **41**, 1767
polycyclic conjugated hydrocarbons	*Tetrahedron* (1993), **49**, 9207
polyepoxides	*Tetrahedron* (1980), **36**, 461
polyepoxides	*Tetrahedron* (1980), **36**, 833
polyfluoroaromatic compounds	*Synthesis* (1976), 652
polyketide natural products	*Tetrahedron* (1977), **33**, 2159
polymer-supported reagents	*Synthesis* (1979), 481
polymerization	*Tetrahedron* (1993), **49**, 8707
polymers	*Synthesis* (1997), 1217
polymers, as protecting groups	*Tetrahedron* (1981), **37**, 663
polymers, functionalized	*Synthesis* (1981), 413
polynitropolycyclic cage molecules	*Tetrahedron* (1988), **44**, 2377
polyols	*Synthesis* (1991), 635
polypeptides	*Tetrahedron* (1995), **51**, 9241
polysaccharide modifications	*Tetrahedron* (1985), **41**, 2957
polysaccharides	*Tetrahedron* (1987), **43**, 2389
potassium channels, modulation	*Synthesis* (1996), 307
prebiotic synthesis	*Tetrahedron* (1984), **40**, 1093
prebiotics	*Tetrahedron* (1997), **53**, 11493
Prelog-Djerassi lactonic acid	*Synthesis* (1991), 245
pressure, high, synthesis under	*Synthesis* (1985), 1
pressure, high, synthesis under	*Synthesis* (1985), 999
Prins reaction	*Synthesis* (1977), 661
proline	*Tetrahedron* (1996), **52**, 4527
propandiol, 2-methyl-1, 3-	*Synthesis* (1993), 1029
propargylic compds, hydrogenolysis	*Synthesis* (1996), 1
propargylic silanes in intramolecular addition reactions	*Synthesis* (1988), 263
propionic acids, 2-aryl,	*Tetrahedron* (1986), **42**, 4095
prostaglandin, synthesis	*Tetrahedron* (1980), **36**, 2163
prostanoids	*Synthesis* (1984), 449
prostanoids, natural source	*Tetrahedron* (1995), **51**, 4571
protecting groups	*Synthesis* (1979), 921
protecting groups	*Synthesis* (1980), 1
protecting groups	*Tetrahedron* (1981), **37**, 663
protecting groups, for alcohols	*Synthesis* (1993), 11
protecting groups, allylic	*Synthesis* (1996), 1
protecting groups, allylic	*Tetrahedron* (1997), **53**, 13509
protective groups, carboxylic acids	*Tetrahedron* (1980), **36**, 2409
protective groups, organosilicon	*Synthesis* (1985), 817
protecting groups, photolabile	*Synthesis* (1980), 1

protein overproduction	*Tetrahedron* (1991), **47**, 2543
protein biosynthesis, tRNA	*Tetrahedron* (1977), **33**, 1671
proteinases, cysteine	*Tetrahedron* (1976), **32**, 291
proteins, copper containing	*Tetrahedron* (1989), **45**, 3
protosterols, meralonic acid	*Tetrahedron* (1986), **42**, 3
Pummerer reaction	*Synthesis* (1997), 1353
pyrasolines, 2-	*Synthesis* (1985), 1028
pyrazolidinediones, 3,5-	*Synthesis* (1985), 1028
pyridines	*Synthesis* (1972), 464
pyridines,	*Synthesis* (1976), 1
pyridines, ring reactions	*Tetrahedron* (1981), **37**, 3423
pyridinium compounds	*Tetrahedron* (1980), **36**, 679
pyridones 2-	*Tetrahedron* (1992), **48**, 9111
pyrido[4,3-b]carbazoles	*Synthesis* (1977), 437
pyrimidines, pyrrolo-	*Synthesis* (1974), 837
pyrones	*Tetrahedron* (1977), **33**, 3183
pyrones	*Tetrahedron* (1992), **48**, 9111
pyrrolenines	*Synthesis* (1976), 281
pyrroles	*Synthesis* (1976), 281
pyrroles, 3-substituted	*Synthesis* (1985), 353
pyrroles, thieno-	*Synthesis* (1985), 143
pyrrolizine	*Synthesis* (1987), 10
pyrrolopyrimidines	*Synthesis* (1974), 837
pyrylium cations	*Tetrahedron* (1980), **36**, 679

Q

quadratic acid	*Synthesis* (1978), 869
quaternary ammonium compounds	*Synthesis* (1973), 441
quaternary carbon,	*Tetrahedron* (1991), **47**, 9503
quaternary carbon construction	*Tetrahedron* (1980), **36**, 419
quinodimethanes	*Synthesis* (1978), 793
quinodimethanes	*Tetrahedron* (1987), **43**, 2873
quinododimethide	*Tetrahedron* (1987), **43**, 2873
quinones, nucleophilic reactions of	*Tetrahedron* (1991), **47**, 8043

R

racemization, photochemical	*Tetrahedron* (1997), **53**, 9417
radical cations, dimerization of olefins	*Synthesis* (1974), 539
radical cations, fragmentation of	*Tetrahedron* (1994), **50**, 575
radical chain reactions	*Synthesis* (1988), 417

radical chain reactions — *Synthesis* (1988), 489

radical cyclization — *Tetrahedron* (1997), **53**, 17543

radical, ion reactions — *Tetrahedron* (1985), **41**, 2771

radical reactions — *Synthesis* (1973), 1

radical reactions — *Tetrahedron* (1981), **37**, 3073

radical reactions — *Tetrahedron* (1982), **38**, 313

radical reactions — *Tetrahedron* (1987), **43**, 3541

radical reaction kinetics — *Tetrahedron* (1993), **49**, 1151

radicals — *Tetrahedron* (1996), **52**, 13265

radicals, addition reactions — *Tetrahedron* (1980), **36**, 701

radicals, amidyl — *Tetrahedron* (1978), **34**, 3241

radicals, cyclopropyl — *Tetrahedron* (1981), **37**, 1625

radicals, generation of — *Tetrahedron* (1994), **50**, 575

radicals, in alcohol deoxygenation — *Tetrahedron* (1983), **39**, 2609

radicals, short contact time reactions — *Tetrahedron* (1986), **42**, 2135

radicals, stereochemistry — *Tetrahedron* (1980), **36**, 2531

reactivity, effect of strain — *Tetrahedron* (1985), **41**, 1613

rearrangement, Claisen — *Synthesis* (1977), 589

rearrangements, carbocyclic spiros — *Synthesis* (1976), 425

rearrangements, circumambulatory — *Tetrahedron* (1982), **38**, 567

rearrangements in organoborons — *Synthesis* (1973), 635

rearrangements in LAH reductions — *Synthesis* (1974), 691

rearrangements, silver ion catalyzed — *Synthesis* (1975), 347

redox radical reactions — *Tetrahedron* (1995), **51**, 7579

reduction, Birch conditions — *Synthesis* (1972), 391

reduction, by low valent transition metal complexes — *Tetrahedron* (1988), **44**, 4295

reduction by ionic hydrogenation — *Synthesis* (1974), 633

reduction, catalytic hydrogen transfer — *Synthesis* (1988), 91

reduction, homogeneous asymmetric — *Synthesis* (1981), 85

reduction, hydride — *Tetrahedron* (1979), 35, 567

reduction, hydrogen transfer — *Synthesis* (1988), 91

reduction, tri-*n*-butyltin hydride — *Synthesis* (1987), 665

reduction, with alkali metals — *Tetrahedron* (1986), **42**, 6351

reduction with alkoxyaluminohydrides — *Synthesis* (1972), 217

reduction, with Dibal — *Synthesis* (1975), 617

reduction, with group IV–VIB metals — *Synthesis* (1979), 1

reduction with lithium/amine — *Synthesis* (1972), 391

reduction, with LAH	*Synthesis* (1974), 691
reduction with sodium cyanoborohydride	*Synthesis* (1975), 135
reduction with sulfurated borohydrides	*Synthesis* (1972), 526
reduction with thexylborane	*Synthesis* (1974), 77
reduction, with Tibal	*Synthesis* (1975), 617
reductive cleavage of aliphatic nitros	*Synthesis* (1986), 693
reductones	*Synthesis* (1972), 176
resolution, optical strategies	*Tetrahedron* (1977), **33**, 2725
ring enlargement, erocycles	*Synthesis* (1993), 931
resinsulfonic acid Nafion-H	*Synthesis* (1986), 513
retro Diels-Alder reaction	*Synthesis* (1977), 270
retro-Diels-Alder reaction	*Synthesis* (1987), 207
retro-ene reaction	*Synthesis* (1993), 659
rhodium compounds	*Tetrahedron* (1991), **47**, 1765
ribonucleotides	*Tetrahedron* (1993), **49**, 10441
ring closures, heterocyclic	*Tetrahedron* (1987), **43**, 5171
ring construction by the Robinson annelation	*Synthesis* (1976), 777
ring degeneration, of azines	*Tetrahedron* (1985), **41**, 237
ring enlargement, for macrocyclics	*Tetrahedron* (1988), **44**, 1573
ring expansion, medium rings by	*Tetrahedron* (1993), **49**, 10749
ring expansions of bridged bicyclic ketones	*Tetrahedron* (1987), **43**, 3
ring, four-membered from isocyanides	*Synthesis* (1985), 1083
ring transformations of isoxazoles	*Synthesis* (1975), 20
RNA, transfer in protein synthesis	*Tetrahedron* (1977), **33**, 1671
Robinson annelation	*Synthesis* (1976), 777
Robinson annelation	*Tetrahedron* (1976), **32**, 3

S

Sandmeyer reaction	*Synthesis* (1988), 923
α-santonin, in terpene synthesis	*Tetrahedron* (1993), **49**, 4761
Schiff bases, reaction with dienophiles	*Synthesis* (1976), 349
selenium compounds	*Tetrahedron* (1978), 34, 1049
selenium compounds, extrusions	*Tetrahedron* (1988), **44**, 6241
selenium compounds, selenoethers	*Synthesis* (1988), 749

selenium compounds, tetraselenafulvalenes — *Synthesis* (1976), 489

seleno-modified nucleosides — *Tetrahedron* (1993), **49**, 9877

selenoethers, dealkylation — *Synthesis* (1988), 749

semiempirical methods — *Tetrahedron* (1988), **44**, 7393

sesquiterpenes, bioconversion of — *Tetrahedron* (1990), **46**, 4109

sesquiterpenes, polycarbocyclic — *Tetrahedron* (1985), **41**, 1767

sesquiterpenoids, natural source — *Tetrahedron* (1995), **51**, 4571

seven-membered heterocycles via pericyclic reactions — *Synthesis* (1988), 569

seven-membered rings — *Tetrahedron* (1993), **49**, 5203

Sharpless epoxidation — *Synthesis* (1986), 89

shikimic acid pathway — *Tetrahedron* (1978), **34**, 3353

shikimic acid — *Synthesis* (1993), 179

sialic acid, glycosidation of — *Tetrahedron* (1990), **46**, 5835

silanes, allylic and propargylic — *Synthesis* (1988), 263

silanes, intramolecular additions — *Synthesis* (1988), 263

silanes, chloromethyltrimethylsilane — *Synthesis* (1985), 717

silated enolates — *Synthesis* (1977), 91

silicon compounds — *Synthesis* (1979), 761

silicon compounds — *Synthesis* (1980), 861

silicon compounds — *Tetrahedron* (1988), **44**, 2675

silicon compounds — *Synthesis* (1993), 349

silicon compounds, aryltrimethyl — *Synthesis* (1979), 841

silicon compounds, intramolecular additions — *Synthesis* (1988), 263

silicon compounds, organosilicon protecting groups — *Synthesis* (1985), 817

silicon compounds, Peterson reaction — *Synthesis* (1984), 384

silicon compounds, silylation methods — *Synthesis* (1982), 1

silicon compounds, with sulfur — *Tetrahedron* (1988), **44**, 281

silicon ethers — *Synthesis* (1997), 813

silicon, interaction with positive carbon — *Tetrahedron* (1990), **46**, 2677

silicon linkages — *Synthesis* (1997), 813

silicon reagents — *Synthesis* (1980), 861

siloxy dienes — *Tetrahedron* (1995), **51**, 4571

silver ion catalyzed rearrangements — *Synthesis* (1975), 347

silyl dienes — *Synthesis* (1993), 349

silyl enol ethers — *Synthesis* (1983), 1

silyl enol ethers — *Synthesis* (1983), 85

single electron transfer reactions   *Tetrahedron* (1995), **51**, 7579

singlet oxygen   *Tetrahedron* (1981), **37**, 1825

sodium perborate   *Synthesis* (1995), 1325

sodium perborate, in oxidations   *Tetrahedron* (1995), **51**, 6145

sodium percarbonate, in oxidations   *Tetrahedron* (1995), **51**, 6145

sodium percarbonate   *Synthesis* (1995), 1325

sodium cyanoborohydride   *Synthesis* (1975), 135

solid state, organic reactivity   *Tetrahedron* (1994), **50**, 6441

solid phase reagents   *Synthesis* (1979), 40

solid phase reagents   *Tetrahedron* (1979), 35, 723

solid phase synthesis   *Tetrahedron* (1993), **49**, 11065

solid-phase fragment condensation, in peptide synthesis   *Synthesis* (1994), 337

solvent effects in 1, 3-dipolar cycloaddition reactions   *Synthesis* (1973), 71

solvolysis reactions, gas phase   *Tetrahedron* (1982), **38**, 3195

spin labeling, phosphorus   *Synthesis* (1981), 682

spiro compound   *Synthesis* (1978), 77

spiro compounds by intramolecular alkylation   *Synthesis* (1974), 383

spiro compounds, [2.2]phanes   *Synthesis* (1973), 85

spiro compounds via rearrangements   *Synthesis* (1976), 425

spirobenzopyran, for membranes   *Tetrahedron* (1994), **50**, 4039

spiroketal subunits in polyethers   *Tetrahedron* (1987), **43**, 3309

squaric acid, reactions   *Synthesis* (1980), 961

squaric acid   *Synthesis* (1978), 869

$S_{RN}1$-$S_{RN}2$ mechanisms   *Tetrahedron* (1993), **49**, 4485

stacking effects   *Synthesis* (1995), 475

stannyllithiums   *Synthesis* (1990), 259

Staudinger reaction   *Tetrahedron* (1992), **48**, 1353

Stenhouse salts, use   *Tetrahedron* (1977), **33**, 463

stereochemistry, alkenes   *Tetrahedron* (1980), **36**, 557

stereochemistry, allylic pyrophosphate metabolism   *Tetrahedron* (1980), **36**, 1109

stereochemistry, asymmetric synthesis   *Tetrahedron* (1979), **35**, 2797

stereochemistry, asymmetric synthesis   *Tetrahedron* (1980), **36**, 2

stereochemistry, cyclohexenones   *Tetrahedron* (1982), **38**, 3

stereochemistry, enzymatic reactions   *Tetrahedron* (1982), **38**, 1541

stereochemistry, ketone reduction   *Tetrahedron* (1979), **35**, 449 stereochemistry, phosphorus — *Tetrahedron* (1980), **36**, 2059

stereochemistry, topological — *Tetrahedron* (1985), **41**, 3161

stereochemistry, torsion angle rotation — *Tetrahedron* (1980), **36**, 2809

stereochemistry, sulfenamides and hydroxylamines — *Tetrahedron* (1984), **40**, 3345

stereocontrolled reactions, of organometallics — *Tetrahedron* (1993), **49**, 10175

stereoselective synthesis of amino acids — *Tetrahedron* (1994), **50**, 1539

sterically crowded molecules — *Tetrahedron* (1978), **34**, 1855

steroidal ketones, reduction with sulfurated borohydrides — *Synthesis* (1972), 526

steroids, total synthesis — *Tetrahedron* (1981), **37**, 3

steroids, C-nor-D-homo — *Tetrahedron* (1979), 35, 911

sterols, meralonic acid — *Tetrahedron* (1986), **42**, 3

stilbenes — *Synthesis* (1983), 341

stilbides, synthesis and reactions — *Synthesis* (1974), 328

strain, relation to reactivity — *Tetrahedron* (1985), **41**, 1613

strained molecules — *Tetrahedron* (1978), **34**, 1855

structure determination, — *Tetrahedron* (1991), **47**, 3521

substitution, aromatic, of olefins by palladium salts — *Synthesis* (1973), 524

substitution, electrophilic diazoalkane — *Synthesis* (1985), 569

substitution, H in azines — *Tetrahedron* (1988), **44**, 1

substitution reactions, anodic — *Tetrahedron* (1976), **32**, 2185

substitutive acyloxylation at carbon — *Synthesis* (1972), 1

succinaldehyde, monoacetal — *Synthesis* (1985), 592

sulfates, cyclic — *Synthesis* (1992), 1035

sulfenes thione dioxides, perhalogenated — *Synthesis* (1988), 349

sulfides, α-chloro, synthesis with — *Tetrahedron* (1986), **42**, 3731

sulfides, cyclic — *Tetrahedron* (1982), **38**, 2857

sulfines thione oxides, perhalogenated — *Synthesis* (1988), 349

sulfinimines, Ei reaction — *Tetrahedron* (1977), **33**, 2359

sulfites, cyclic — *Synthesis* (1992), 1035

sulfones — *Tetrahedron* (1977), **33**, 2019

sulfonium salts, halo — *Tetrahedron* (1982), **38**, 2597

sulfoxides, chiral — *Synthesis* (1981), 185

sulfoxides, from thioethers — *Tetrahedron* (1988), **44**, 6537

sulfoxides, from thioethers — *Tetrahedron* (1986), **42**, 5459

sulfoxonium, dimethyl methylide   *Tetrahedron* (1987), **43**, 2609
sulfur compounds   *Synthesis* (1972), 101
sulfur compounds   *Tetrahedron* (1991), **47**, 1109
sulfur compounds   *Tetrahedron* (1989), **45**, 7643
sulfur compounds   *Synthesis* (1978), 713
sulfur compounds   *Synthesis* (1992), 1035
sulfur compounds   *Synthesis* (1978), 713
sulfur compounds   *Tetrahedron* (1988), **44**, 6241
sulfur compounds   *Tetrahedron* (1988), **44**, 6537
sulfur compounds, sbiosynthesis   *Tetrahedron* (1983), **39**, 1215
sulfur compounds, α-chlorosulfides   *Tetrahedron* (1986), **42**, 3731
sulfur compounds, in cycloadditions   *Tetrahedron* (1988), **44**, 6755
sulfur compounds by   *Synthesis* (1985), 586
    dehalogenation with sulfur
sulfur compounds,   *Tetrahedron* (1987), **43**, 2609
    dimethylsulfoxonium methylide
sulfur compounds, dithio acids   *Synthesis* (1983), 605
sulfur compounds, ethoxy carbonyl   *Synthesis* (1975), 301
    isothiocyanate
sulfur compounds, extrusions   *Tetrahedron* (1988), **44**, 6241
sulfur compounds, heterocyclic   *Tetrahedron* (1982), **38**, 3537
sulfur compounds, by Lawesson's   *Tetrahedron* (1985), **41**, 5061
    reagents
sulfur compounds, α-oxoketene   *Tetrahedron* (1986), **42**, 3029
    dithioacetals
sulfur compounds, perhalogenated   *Synthesis* (1988), 349
    sulfenes
sulfur compounds, perhalogenated   *Synthesis* (1988), 349
    sulfines
sulfur compounds, with silicon   *Tetrahedron* (1988), **44**, 281
sulfur compounds, from thioethers   *Tetrahedron* (1986), **42**, 5459
sulfur compounds, from thioethers   *Tetrahedron* (1988), **44**, 6537
sulfur compounds, sultones   *Tetrahedron* (1987), **43**, 1027
sulfur compounds,   *Synthesis* (1976), 489
    tetrathiafulvalenes
sulfur compounds, 1,3-thiazines   *Synthesis* (1972), 333
sulfur compounds, thienopyrroles,   *Synthesis* (1985), 143
sulfur compounds, α-thioalkylation   *Synthesis* (1987), 589
sulfur compounds, thioethers,   *Synthesis* (1988), 749
    selective dealkylation
sulfur compounds, thioketenes   *Tetrahedron* (1988), **44**, 1827
sulfur compounds, thiolactones   *Synthesis* (1972), 151

sulfur compounds, thiocarbonyl, photochemistry — *Tetrahedron* (1985), **41**, 5393

sulfur compounds, thiophosgene — *Synthesis* (1978), 803

sulfur in synthesis — *Tetrahedron* (1988), **44**, 6241

sulfur-modified nucleosides — *Tetrahedron* (1993), **49**, 9877

sulfur-silicon compounds, mixed — *Tetrahedron* (1988), **44**, 281

sulfur ylides polarization reversal — *Synthesis* (1977), 357

sulfurated borohydrides — *Synthesis* (1972), 526

sultones — *Tetrahedron* (1987), **43**, 1027

supported reagents — *Synthesis* (1979), 40

supported reagents — *Synthesis* (1997), 1217

supramolecular chemistry — *Tetrahedron* (1997), **53**, 15911

supramolecules, third generation — *Tetrahedron* (1993), **49**, 8933

synthesis, asymmetric — *Synthesis* (1978), 329

synthesis, asymmetric — *Tetrahedron* (1986), **42**, 5157

synthesis, asymmetric — *Synthesis* (1990), 541

synthesis, asymmetric — *Tetrahedron* (1990), **46**, 5029

synthesis, asymmetric — *Synthesis* (1991), 1

synthesis, asymmetric — *Synthesis* (1991), 103

synthesis, asymmetric — *Tetrahedron* (1991), **47**, 6079

synthesis by ring enlargement — *Tetrahedron* (1988), **44**, 1573

synthesis, large scale — *Tetrahedron* (1991), **47**, 4789

synthesis by reductive cleavage of nitro groups — *Synthesis* (1986), 693

synthesis, stereocontrol — *Tetrahedron* (1980), **36**, 2

synthesis, stereoselective — *Synthesis* (1991), 635

synthesis, steric crowding in — *Tetrahedron* (1978), **34**, 1855

synthetic building blocks — *Synthesis* (1987), 1

synthons, bicyclo[3.2.0]heptanones — *Synthesis* (1977), 155

synthons, epoxide-like — *Synthesis* (1992), 1035

synthesis with azomethine ylids — *Synthesis* (1973), 469

T

tandem reactions — *Tetrahedron* (1995), **51**, 13103

tellurium compounds, extrusion reactions — *Tetrahedron* (1988), **44**, 6241

tellurium reagents — *Synthesis* (1986), 1

tellurium reagents — *Synthesis* (1991), 793

tellurium reagents — *Synthesis* (1991), 897

telomerization, bromo addends — *Synthesis* (1977), 145

terpenes, metabolism in mammals — *Tetrahedron* (1984), **40**, 3597

terpenoid compounds   *Tetrahedron* (1993), **49**, 4761
terpenoids, in basidiomycetes   *Tetrahedron* (1981), **37**, 2199
tetracyanoethylene   *Synthesis* (1986), 249
tetracyanoethylene   *Synthesis* (1987), 749
tetracyanoethylene, with   *Synthesis* (1987), 959
   organometallics
tetraheterofulvalenes   *Tetrahedron* (1986), **42**, 1209
tetrahydrofuran subunits in   *Tetrahedron* (1987), **43**, 3309
   polyethers
tetrahydrofurans, 2,5-disubstituted   *Synthesis* (1995), 1447
tetrahydropyran units in polyethers   *Tetrahedron* (1987), **43**, 3309
tetraselenafulvalenes, preparation   *Synthesis* (1976), 489
tetrathiafulvalenes TTF, preparation   *Synthesis* (1976), 489
tetrazols, azido-, isomerization   *Synthesis* (1973), 123
thermal isomerization   *Tetrahedron* (1976), **32**, 641
thermal reactions, asymmetric   *Tetrahedron* (1993), **49**, 293
thermolytic reactions,   *Synthesis* (1976), 374
   fluoro-organics
thexylborane   *Synthesis* (1974), 77
thiazines   *Synthesis* (1972), 333
thiazines   *Synthesis* (1985), 586
thiazoles, hydroxy   *Tetrahedron* (1980), **36**, 2023
thiazolothiazole   *Synthesis* (1985), 586
thienopyrroles   *Synthesis* (1985), 143
thioalkylation   *Synthesis* (1987), 589
thioation reactions with Lawesson's   *Tetrahedron* (1985), **41**, 5061
   reagents
thiocarbonyl ylides   *Tetrahedron* (1976), **32**, 2165
thiocarbonyl compounds,   *Tetrahedron* (1985), **41**, 5393
   photochemistry
thiocarbonyl compounds   *Synthesis* (1992), 1185
thioethers, dealkylation   *Synthesis* (1988), 749
thioethers, oxidation to sulfoxides   *Tetrahedron* (1986), **42**, 5459
thioethers, oxidation to sulfoxides   *Tetrahedron* (1988), **44**, 6537
thioisomunchnones, cycloaddition of   *Synthesis* (1994), 123
thioketenes   *Tetrahedron* (1988), **44**, 1827
thiolactones, heterocycles   *Synthesis* (1972), 151
thiols, thioalkylation   *Synthesis* (1987), 589
thione dioxides sulfenes,   *Synthesis* (1988), 349
   perhalogenated
thione oxides sulfines,   *Synthesis* (1988), 349
   perhalogenated
thionyl chloride   *Synthesis* (1981), 661

thiophosgene	*Synthesis* (1978), 803
thiophosphorus, reagents	*Tetrahedron* (1985), **41**, 2567
transition metal complexes, additions	*Tetrahedron* (1978), **34**, 3047
tin compounds	*Synthesis* (1990), 259
tin compounds	*Synthesis* (1992), 803
tin compounds	*Tetrahedron* (1993), **49**, 7395
tin compounds, hydroxyl manipulation	*Tetrahedron* (1985), **41**, 643
tin compounds, tributyltin hydride	*Synthesis* (1987), 665
tin hydride in radical reactions	*Synthesis* (1988), 489
tin hydride, tri-*n*-butyl	*Synthesis* (1987), 665
tin hydride in radical reactions	*Synthesis* (1988), 417
topological stereochemistry	*Tetrahedron* (1985), **41**, 3161
transition metal complexes in diynes	*Synthesis* (1974), 761
transition metal catalyst in carbonylation reactions	*Synthesis* (1973), 509
transition metal catalyst, optically active	*Synthesis* (1988), 645
transition metal complexes, low valent, reduction by	*Tetrahedron* (1988), **44**, 4295
transition metals	*Synthesis* (1976), 561
transmetalation	*Synthesis* (1993), 537
tri-*n*-butyltin hydride	*Synthesis* (1987), 665
tricarbonyl compounds, cyclic	*Tetrahedron* (1978), **34**, 1285
trichloromethyl reduction	*Synthesis* (1983), 773
triflates, vinyl and aryl	*Synthesis* (1993), 735
trifluoromethylation	*Tetrahedron* (1992), **48**, 6555
triisobutylaluminium hydride,	*Synthesis* (1975), 617
trithiadiazapentalenee	*Synthesis* (1985), 586

U

Ullmann synthesis of biaryls	*Synthesis* (1974), 9
umplung of carbonyl reactivity	*Synthesis* (1977), 357
unsaturated carbonyl compounds, cyclization	*Synthesis* (1975), 1
unsaturated compounds, reactions	*Synthesis* (1986), 793
unsaturated ketones, oxidation	*Synthesis* (1992), 235
unsaturated systems, and manganese	*Synthesis* (1993), 833
ureas, cyclic by α-ureidoalkylation	*Synthesis* (1973), 243
uroporphyrinogen III synthase	*Tetrahedron* (1991), **47**, 6003

V

vinyl bromides, reactions	*Synthesis* (1995), 745
vinyl carbanions, reactions	*Tetrahedron* (1988), **44**, 4653
vinyl stannanes, fluorination	*Tetrahedron* (1995), **51**, 6605
vinyl sulphones, chemistry of	*Tetrahedron* (1990), **46**, 6951
vinylene triheterocarbonates	*Tetrahedron* (1986), **42**, 1209
vinylphosphine oxides	*Synthesis* (1974), 775
vinylphosphonates	*Synthesis* (1992), 333
vinylphosphonium salts	*Synthesis* (1974), 775
vitamin A, stereoisomers	*Tetrahedron* (1984), **40**, 1931
vitamin B_{12}	*Tetrahedron* (1975), **31**, 2639

W

Weiss reaction	*Tetrahedron* (1991), **47**, 3665
Willgerodt reaction	*Synthesis* (1975), 358
Wittig reaction, intramolecular	*Tetrahedron* (1980), **36**, 1717
Wittig reactions, *bis-*	*Synthesis* (1975), 765
Wittig sigmatropic rearrangement	*Synthesis* (1991), 594

X

xanthonoids, plant	*Tetrahedron* (1980), **36**, 1465
o-xylylene	*Tetrahedron* (1987), **43**, 2873

Y

ylidenemalonodinitriles, cyclizations	*Synthesis* (1976), 705
ylids, azomethine,	*Synthesis* (1973), 469
ylides, carbonyl	*Tetrahedron* (1976), **32**, 2165
ylides, cyclicimmonium	*Tetrahedron* (1976), **32**, 2647
ylides, sulfur	*Synthesis* (1977), 357
ynamine	*Tetrahedron* (1976), **32**, 1449

Z

zinc compounds, activation methods	*Tetrahedron* (1987), **43**, 2203
zinc reagents	*Tetrahedron* (1992), **48**, 9577
zirconium compounds	*Synthesis* (1988), 1
zirconocene compounds	*Tetrahedron* (1995), **51**, 4255

INDEX

A

Aldrich Catalog/Handbook of Fine Chemicals, 1–3
Annual Reports in Organic Synthesis, 5–10

B

Beilstein's Handbook of Organic Chemistry, 11–49
abbreviations, 20–24
distribution of compound types, 16–19
list of prefixes, 28–49
overview, 11–15
rules of the Beilstein System of Indexing, 25–27

C

Chemical Abstracts, 51–81
Chemical Abstracts sections, 74–81
online file names, 59–71
overview, 51–56
patent country codes, 72–73
roles, 57–59
Chemist's Companion, 83–90
Chemistry of Functional Groups Series (Patai's), 229–242
Compendium of Organic Synthesis Methods, 91–94
Comprehensive Organic Chemistry, 95–98
Comprehensive Organic Functional Group Transformations, 107–108
Comprehensive Organic Transformations: A Guide to Functional Group Preparations, 109–110
Comprehensive Organometallic Chemistry, 99–106
CRC Handbook of Chemistry and Physics, 111–135
CRC Handbook of Data on Organic Compounds, 137–141

D

Dictionaries, scientific, 287–288
Dictionary of Organic Compounds (Heilbron), 147–150
Dictionary of Organometallic Compounds, 143–146
Dissertations, 151–152

E

Encyclopedia of Reagents for Organic Synthesis, 153–158

F

Fieser's Reagents for Organic Synthesis, 251–252

Formation of C–C Bonds
    (Mathieu), 159–166

H
Heilbron Dictionary of Organic
    Compounds, 147–150
Houben-Weyl Methoden der
    Organischen Chemie, 167–178

I
Internet resources, 179–180

J
Journals, 181–185

K
Kirk-Othmer Encyclopedia of
    Chemical Technology,
    187–188

L
Lange's Handbook of Chemistry,
    189–214
Literature sources, 215–216

M
Mathieu, Formation of C–C
    Bonds, 159–166
Merck Index, 217–222

O
Organic chemistry reviews,
    309–355
Organic Reactions, 223
Organic Synthesis, 225–228

P
Patai's Chemistry of Functional
    Groups Series, 229–242
Patents, 243–248
Purchasing Chemicals, 249–250

R
Reagents for Organic Synthesis
    (Fieser's), 251–252
Rodd's Chemistry of Carbon
    Compounds, 253–282

S
Science Citation Index, 283–286
Scientific Dictionaries, 287–288

T
Theilheimer's Synthetic Methods
    of Organic Chemistry,
    289–303
    functional group
        transformations, 303
    index to systematic
        classification, 293–302
    overview, 289–292
Translations, 305–309

Printed in the United States
3158